MONOCLONAL ANTIBODIES
THE HOPEFUL DRUGS

MONOCLONAL ANTIBODIES
THE HOPEFUL DRUGS

Fathimunisa Begum

Reader
Department of Zoology
Holy Cross College
Tiruchirapalli
Tamil Nadu

MJP PUBLISHERS

ISBN 10: 81-8094-048-9
ISBN 13: 978-81-8094-048-4

Cataloguing-in-Publication Data

Fathimunisa Begum (1951 –).
Monoclonal Antibodies—the Hopeful Drugs / by Fathimunisa Begum. – Chennai : MJP Publishers, 2008
xiv, 264 p. ; 23 cm.
Includes Glossary, References and Index.
ISBN 81-8094-048-9 (pbk.)
1. Antibodies, Immunity. I. Title.
616.0798 FAT MJP 047

ISBN 81-8094-048-9

Printed and bound in India

MJP PUBLISHERS
47, Nallathambi Street
Triplicane
Chennai 600 005

Publisher : J.C. Pillai
Managing Editor : C. Sajeesh Kumar
Project Editor : P. Parvath Radha
Acquisitions Editors : C. Janarthanan
Assistant Editors : B. Ramalakshmi, S. Revathi
Composition : Lissy John, N. Yamuna Devi, C. M. Joys Mary
Cover Designer : Lissy John
CIP Data : Prof. K. Hariharan

To

My husband S T Madani
and my loving children
Sajjad, Bathul and Kaneez

Preface

Monoclonal antibodies—the hopeful drugs invokes an awareness of monoclonal antibody drugs used for the treatment of diseases like cancer, autoimmune diseases, cardiovascular diseases, allergies and other diseases caused by infection.

After an introduction to monoclonals in the first chapter, the production of monoclonal antibodies is elaborated in the second chapter, covering steps like immunization, fusion, screening and selection, cloning and preservation of hybridoma cells. The third chapter covers the *in vivo* and *in vitro* production of antibodies on a large scale.

The fourth chapter presents an account of the techniques used to make human monoclonals. These techniques include combinatorial library, phage display library, protein engineering, gene therapy, etc. The fifth chapter concentrates on the methodology of production of monoclonal antibodies by transgenic plants and animals.

The sixth chapter enumerates the wide application of monoclonals in various fields like medical, environmental, industrial and research. Their use, in cancer diagnosis and treatment and as biosensors, is discussed in detail. A list of antibody drugs approved by Food and Drug Administration (FDA) are given.

The last chapter details each of the FDA-approved drugs. The nature of monoclonals, their mode of action and the experiments undertaken during their clinical trials are presented with appropriate diagrams and graphic illustrations. The research and developments undertaken by various biotech companies, the patents applied, and the present and future status of monoclonals in the drug market are furnished in detail.

As monoclonal antibodies and its applications are part of syllabus for all postgraduate biology courses, this book may be useful for postgraduate students and research scholars of zoology, biochemistry, microbiology, and biotechnology. It may be of immense help to the students of medicine and pharmacy. It may also be handy for those who desire to know the recent therapies adopted for various diseases.

For those who are involved in research work in the field of monoclonals, this book provides basic and overall idea of the research carried out in this field for the past three decades.

Fathimunisa Begum

Contents

1

INTRODUCTION TO HYBRIDOMA TECHNOLOGY

"There is no other technology being used more often," says Donald Drakeman, Medarex pharmaceutical company's President and CEO. Richard van den Broek, a biopharmaceutical analyst at Chase H and Q (New York), states that the "driving success of biotechnology for the past few years has been through monoclonal antibodies." A monoclonal antibody against a specific antigen has proved to be an invaluable tool to identify, locate, quantify or destroy wherever appropriate the cell carrying its target. The potential uses of monoclonals are very widespread.

Researchers have been investigating antibodies as potential therapeutic agents for more than a century. One of the biggest advances in the use of antibodies to treat diseases was made in 1975, when Georges Kohler and Cesar Milstein discovered that monoclonals can be produced by hybridoma cells (hybrid cells) that are formed by the fusion of an antibody-secreting murine lymphocyte cell with a murine myeloma cell. With three decades of tireless work of many, the monoclonals have become a reality as human therapeutics. Monoclonals now represent a large, well-established market that continues to grow dramatically each year. Antibodies currently on the market in the United States had global sales of greater than $13 billion in 2005. Growth in this market is being fuelled by continuing increase in the sales of current products, especially the blockbuster products like Rituximab, Herceptin and Avastin. According to Data monitor, the market is expected to more than triple between 2002 and 2008, from $5.4 billion to $16.7 billion.

During the natural immune response or after immunization in the laboratory, an organism generates a mixture of antibodies with different specificities and affinities. This heterogeneous mixture of antibodies results as a response to the different epitopes found on the immunizing antigen. Many different clones of B cells are involved in the production of this mixture. The serum containing this heterogeneous mixture of antibodies is called polyclonal antisera (Figure 1.1).

The production of polyclonal antisera has several disadvantages. They are as follows.

- Antisera should be raised in genetically identical animals by using identical preparations of antigen and using identical immunization protocol.
- Antisera can be produced in limited volume only.

- It requires a large number of immunized animals and repeated animal use.
- It is impossible to use the antisera raised in different animals, at different times and in different conditions, for the identical serological reagents in complex clinical tests.
- It is extremely difficult, costly and time-consuming to prepare a homogeneous, pure and specific antibody against a single epitope from this heterogeneous mixture of antisera.
- Finally even the purified antibody may contain a less important antibody that gives an unexpected cross reaction, causing confusion in experimental results.

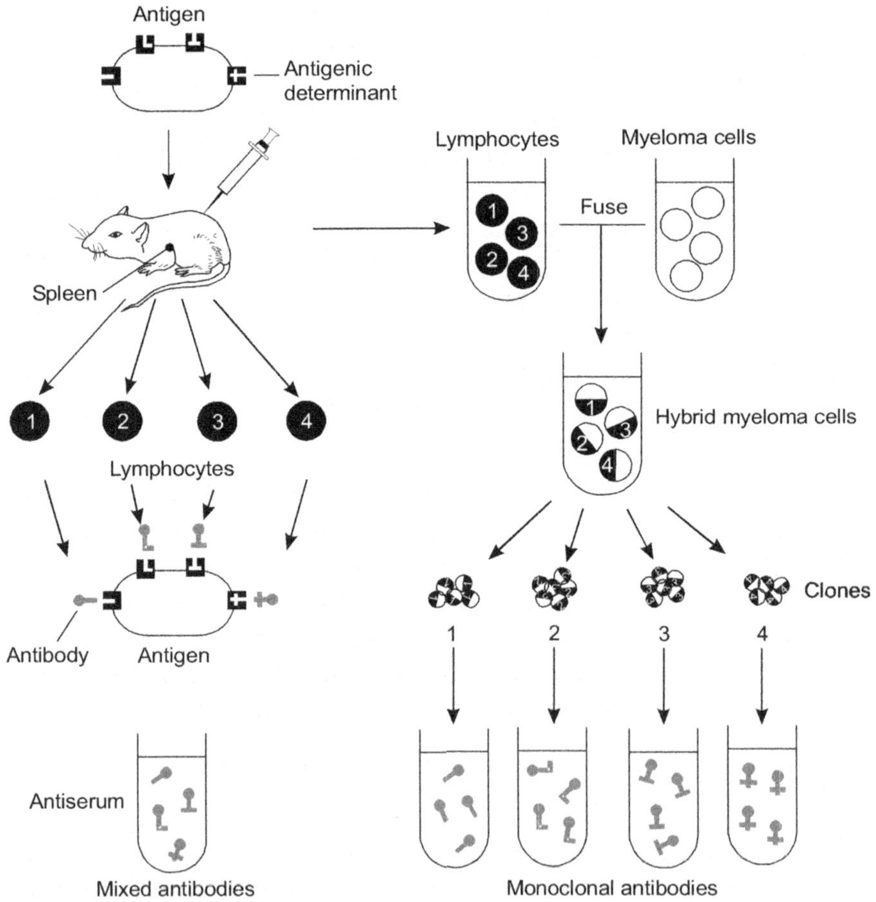

Figure 1.1 Preparation of polyclonal antisera and monoclonal antibodies (Milstein, C., 1991)

2 HYBRIDOMA TECHNOLOGY

INTRODUCTION

Georges Kohler and Cesar Milstein (1975) found that when the normal antibody-secreting B cells are fused with myeloma cells or cancerous B cells, hybridoma cells can be obtained. This fused cell seemed to have the properties of both the fusion products, that is, they secrete antibodies with some antigenic specificity and are also immortal in correct tissue culture conditions. The obtained hybridomas are then cloned and screened for desired antibody specificity. These antibodies are termed as monoclonal antibodies (or MAbs). The nature of antibody depends on the nature of antigen that was injected into the mouse before the B-lymphocyte cells were harvested.

Monoclonal antibodies have several significant advantages compared with polyclonal antiserum.

- They are pure preparations, with known antigenic specificity and affinity.
- They are of a single Ig class.
- They are available in unlimited amount.

The major disadvantages of monoclonal antibodies are

- They possess poor precipitating or agglutinating ability because they bind to only one epitope.
- The production of MAbs is labour-intense, and depends to some extent on luck.
- Repeated screening is needed to test the specificity.

The hybridoma production requires a sequence of steps (Figure 2.1) which are detailed in the following sections.

HYBRIDOMA PREPARATION

Immunization

It is the initial and most crucial aspect of hybridoma production. To date, most hybridoma work utilized B cells from BALB/c strain of mice because of availability of well-characterized myeloma cell line of BALB/c origin. Mice are immunized with an antigen that is prepared for injection either by emulsifying the antigen with Freund's adjuvant or with other adjuvants, or by homogenizing a gel slice that contains the antigen. Intact cells, whole membrane and microorganisms are used as immunogens. The choice of the animal, purity of antigen, dose, route of administration, number of booster injections, length of time interval and use of adjuvants can influence the extent of immune response.

A dose of 1 to 10 million cells of particulate antigen or 5 μg of soluble antigen emulsified in CFA (complete Freund's adjuvant) is injected subcutaneously or intraperitonially into the mouse. The animal is injected, on several occasions, to ensure a high titre of antibody. The final dose of antigen is given intravenously three days before the animal is euthanized. It is necessary to use test bleeds to measure the amount of circulating antibodies in the serum.

When a sufficient antibody titre is reached in the serum, the immunized mice are euthanized, and their spleens are removed (Figure 2.1). This spleen contains the lymphocytic cells for fusion with myeloma cells. If the titre is low, mice can be boosted until an adequate response is achieved, as determined by repeated blood samples. In a batch of 5 to 6 immunized mice, three can be used at a time and the other set can be used later.

The type of antibody produced is also dependent on the number of immunizations. If IgM is required, only one immunization is carried out, as only a primary response is required, whereas IgG antibodies require multiple injections at intervals of approximately 3–4 weeks to allow an effective secondary response.

Preparation of Myeloma Cells

The cell lines used for the preparation of myeloma cells should have the following two properties:

1. It must not be capable of secreting antibodies.
2. It should bear a genetic marker that allows selection of fused cells. Kohler and Milstein used the cell lines defective in HGPRT (hypoxanthine guanine phosphoribosyl transferase) enzyme. Originally myeloma cells were obtained from mineral-oil-induced plasmacytes from BALB/c mice, which were selected in 8-azaguanine so that they lack HGPRT enzyme. Today myeloma cells are obtained from LOU/wsl strain with spontaneously developed ileo-caecal plasmacytomas.

Hybridization or Fusion

Three days prior to fusion, myeloma cells are freshly cultured on fresh 8-azaguanine medium and are maintained in exponential growth (log phase). Prior to fusion they are grown in DEM (dimethylmaleate) or RPMI-1640 medium (developed by Roswell Park Memorial Institute, hence the acronym), with the cells showing high viability and rapid growth.

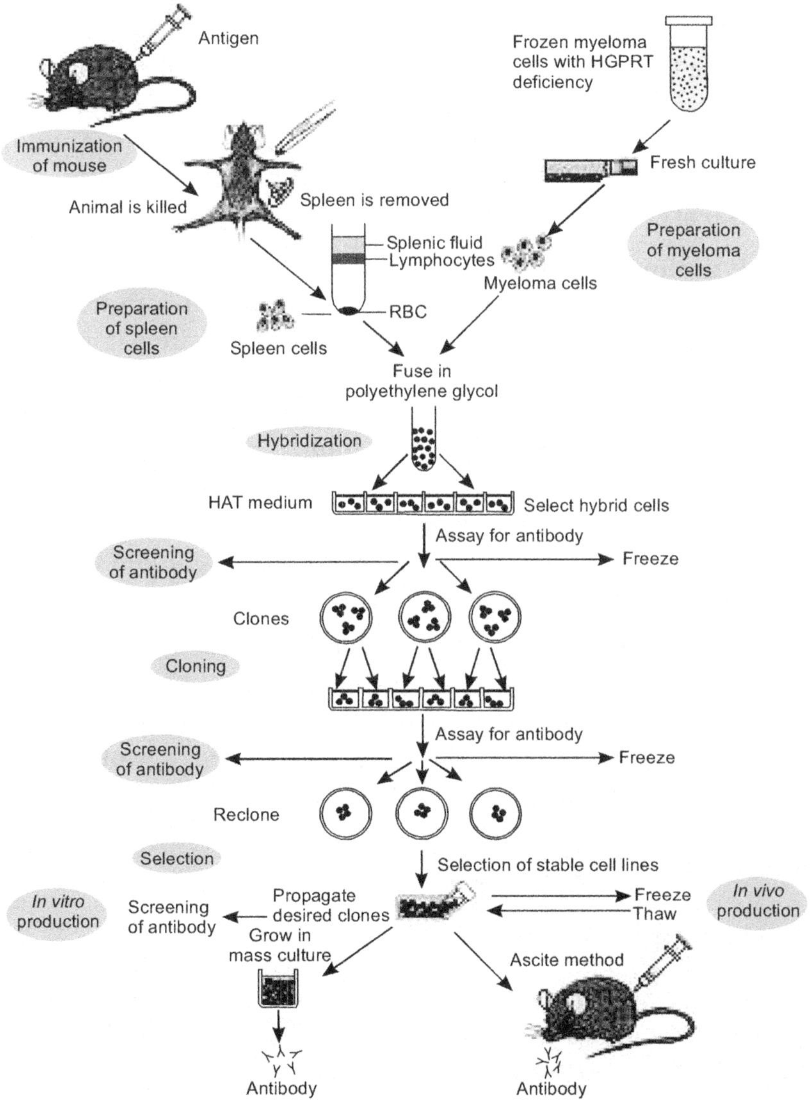

Figure 2.1 Flowchart showing the steps in hybridoma technology (Milstein, C., 1991)

On the day of fusion, they are harvested by pelleting, washed down twice, resuspended in the same medium at 1–2 $\times$ 10^7 cells ml^{-1} concentration, and kept in cold conditions.

Spleens from two or three immunized animals are aseptically removed by making incisions in the abdomen. They are placed in a Petri dish containing 10 ml of 2X Pen-Strep medium and soaked for 5 minutes. The connective tissues are carefully removed and transferred to a new Petri dish with fresh 2X Pen-Strep media at least twice, each time soaking the spleen for 5 minutes. Gently the spleens are minced to achieve a single cell suspension of lymphocytes and red blood cells. By density gradient centrifugation, RBCs are removed from lymphocytes. They are suspended in 2X Pen-Strep medium, washed several times, centrifuged and pelleted.

The lymphocytes are mixed with myeloma cells in a 4 : 1 ratio. The mixture of cells is exposed to a fusion-promoting agent, polyethylene glycol (PEG), only for a few minutes since it is cytotoxic. Initially, only the cell membranes fuse, and two nuclei are present in the cell. Nuclear fusion takes place during mitosis and subsequent generations of the hybridoma cells often result in the loss of some chromosomes. Since cell fusion is a fairly complex process, the success of any particular hybridization depends on the number of variables like healthy spleen cells, log-phase myeloma cells, tissue culture medium and its supplements, and the concentrations of PEG.

The cells are then washed free of PEG by suspending the pellets in diluted fresh DEM or RPMI containing the HAT (hypoxanthine, aminopterin and thymidine) medium. The washed cells comprise a mixture of hybridomas, unfused myeloma cells and lymphocytes. The hybridomas must be selected in HAT medium.

HYBRIDOMA SELECTION

HAT Selection

There are two biosynthetic pathways by which the mammalian cells can produce nucleotides and hence deoxyribonucleic acid (Figure 2.2). They are the de novo pathway and the salvage pathway.

The de novo pathway is the main biosynthetic pathway for nucleic acids. In this pathway, a methyl or formyl group is transferred from an activated form of tetrahydrofolate. This transfer is blocked by aminopterin, a folic acid analogue, affecting the synthesis of nucleic acids.

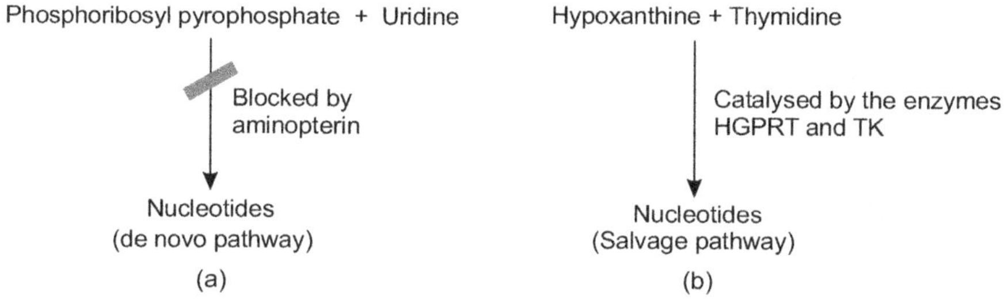

Figure 2.2 Highly simplified diagram showing the two pathways in nucleic acid synthesis (Kuby, J., 1997)

The salvage pathway is catalysed by the enzymes HGPRT and thymidine kinase (TK). A mutation in any one of these enzymes blocks the salvage pathway. The myeloma cells lack the enzyme HGPRT and as a result cannot utilize salvage pathway and are forced to utilize the alternative pathway. However, with the aminopterin drug present in the medium, myeloma cells cannot prepare nucleic acid from this pathway also. Hence they die in the HAT medium.

B cells with functional HGPRT can use both the pathways to synthesize nucleic acids. But in a medium containing hypoxanthine, aminopterin, and thymidine (HAT), unfused myeloma cells cannot grow because they lack the enzyme HGPRT and cannot synthesize DNA in the presence of aminopterin, and unfused normal spleen cells cannot grow indefinitely because of their limited lifespan. However, fused hybridoma cells grow indefinitely, because the spleen cell supplies HGPRT to its myeloma partner, which is immortalized.

HYBRIDOMA CULTURING AND STORAGE

Once the correct hybridoma cells are selected, they can be stored, frozen and cultured whenever required. Since the hybridoma is a transformed cell line, it grows readily in culture but the antibody titre is low.

Culturing Hybridoma Cells

The selected hybridoma cells are then transferred into a 96-well plate with 100–150 μl per well. The cells are grown in a basal medium supplemented with glutamate and buffered with sodium bicarbonate. Feeder cells supplied with growth factors are added to promote the growth of hybridoma cells. The cultures are incubated at 37°C in 5% CO_2 in a CO_2 incubator and examined after 7–10 days for the presence of hybridoma cells. If necessary, fresh HAT medium can be added and incubation is continued.

Assay Development

A plan for screening specific antibody has been done prior to hybridization. Any type of immunoassay can be adopted, i.e., ELISA, RIA, IFA, or immuno flow cytometry. Since the generation time of hybridomas is under 12 hours, short-time assays are the best methods of choice. A known (+) control and a (–) control are needed.

Screening

Not all hybridomas will be producing the antibody of interest. Many will not be producing antibody at all, others will be producing antibodies against the contamination of injected antigen. Screening allows the identification of cells producing the antibody of interest and must be specific for eventual use.

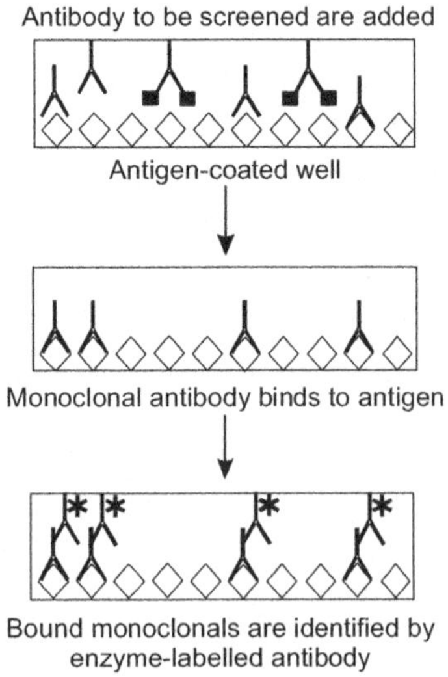

Figure 2.3 Screening of desired monoclonals with ELISA (Dean, C.J., 2003)

The assay most frequently used for hybridoma is ELISA (Figure 2.3.) Both the (–) and (+) controls will need to be included on each assay plate. The (–) control containing all the reagents except the test antibody, gives an indication of the background of the assay. The (+) control containing the antibody is known to react with the bound antigen, producing colour intensity. The purest available antigen must be used for this assay.

In addition, it is desirable to screen for the other properties such as isotype or for specific functional properties such as complement activation or for the ability to elicit ADCC.

Cloning

If all the hybridomas obtained after a fusion were grown together, a polyclonal antibody mixture would be obtained. Consequently, hybridoma cells capable of producing monospecific antibodies need to be isolated by screening. This will identify the wells of microtitre plates producing antibodies of interest. Once identified the contents of these wells should be cloned as early as possible. They are individually grown by diluting a suspension of hybridomas to such an extent that individual aliquots contain an average of only one cell, and by culturing it further in a 24-well plate.

Considerable judgement is necessary at this stage to select the hybridomas capable of expansion versus total loss of cell fusion product due to under-population or inadequate *in vitro* growth at high dilutions. In some instances, the secreted antibody is toxic to the fragile cells maintained *in vitro*. The mouse ascite expansion method at this stage can save the cells. A brief period of growth in mice ascites produces cells that later show enhanced hardiness and optimal antibody production. Following 1–2 weeks of growth, the miniclone is screened for antibody production.

Once suitable hybridomas are identified, they must be cloned several times to ensure a stable, homogeneous colony of cells and to ensure that the antibody produced is indeed monoclonal. Stable positive cells are diluted, reseeded at a density of 1–2 cells for every well in a 6-well plate, allowed to grow, and then screened.

Expansion

Once cloned, colonies are transferred to tissue culture flasks to allow cell growth sufficient for cryopreservation, and/or ascites production. Large-scale expansion can also produce the tissue culture supernatant that contains small amounts of antibody. Hybridomas that have trouble to grow in tissue culture flasks can be cultured in a 24-well plate.

Cryopreservation

To ensure the continuous supply of highly specific reproducible antibody, cell lines are preserved at ultra-low temperatures, while the cells are still being maintained in active culture. The obtained hybridomas are stored frozen for years at liquid nitrogen temperatures: –196°C (liquid) and –151°C (vapour). Only a slight loss of viability is associated in storing them at liquid nitrogen temperatures than storing at temperatures, such as –70°C, which leads to a much more rapid loss of viability. For each cloned cell line, 10–12 vials should be stored, and it should be ensured that at least 3–5 vials are always maintained as stock. DMSO (dimethyl sulphoxide) is the widely used cryopreserving agent and is effective in protecting cells during the freezing and thawing processes.

PRODUCTION OF MONOCLONALS

In vitro Production

Hybridoma cells produce antibodies at a level of 5–25 μg/ml in culture fluid. Cryopreserved cells are expanded in flasks, and cells are removed by centrifugation and filtration. The spent media will contain the antibody. It has contamination of protein from the foetal bovine serum from the media and hence is purified before use.

Ascites Production

Much larger quantities of monoclonals can be obtained by inoculating syngenetic mice with hybridoma cells. The cells grow and produce 100–1000-fold higher titres of antibody (10–60 μg/ml in serum and 1–10 mg/ml in body fluid) and sometimes titres reaching up to 10^6-fold. Mice with ascitic fluids will swell after 7–10 days of inoculation (Figure 2.4). Fluids can be drained 2–3 times from each animal with yields of 3–10 ml each time; however, care must be taken to wait long enough between taps, since death of animal can result.

Fresh ascitic fluids will be straw to red in colour and will contain lipids and also endogenous non-specific mouse antibodies. Ascitic fluids are then centrifuged and the clear supernatant is directly filtered through a sterile membrane.

Figure 2.4 Mouse showing swollen abdomen typical of ascites (John McArdle, 1998, *Animal Welfare Information Center Newsletter*)

A computer database is maintained so that samples can be identified and located quickly.

NATURE OF MONOCLONALS

The most common type of monoclonal is monospecific, bivalent belonging to IgG class of immunoglobulins. Monoclonals of IgM class as well as other much less frequent classes (IgA, IgD and IgE) have also been described. Among these monoclonals, IgM has been particularly used in agglutinating assays and blood typing.

ALTERNATIVE TECHNIQUES FOR HUMAN MONOCLONALS

With the advent of hybridoma, a biotechnological tool, a new era in immunology has commenced. The technique has undergone a tremendous revolution during the past three decades, and the production of human monoclonals have been attempted, as mouse monoclonals are immunogenic for human therapeutic use. These techniques are discussed elaborately in Chapters 3 and 4 and briefly listed (Figure 2.5) in the following section.

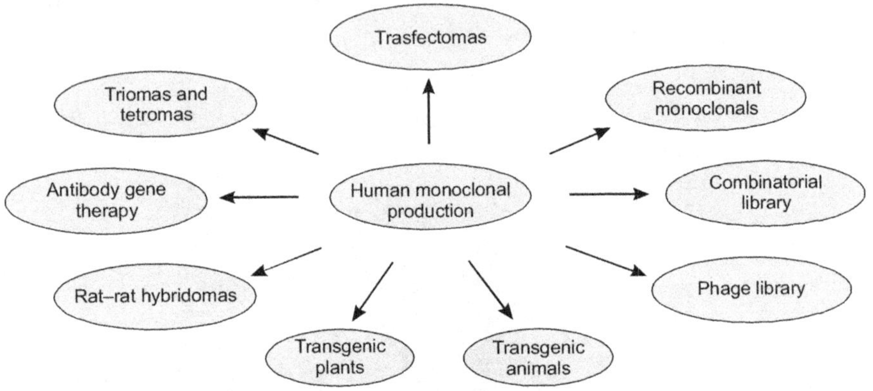

Figure 2.5 Alternative technologies for production of human monoclonals

- The triomas and tetromas (quadromas) can be obtained by fusing hybridoma cells with another spleen cell or with another hybridoma cell.
- Transfectomas are obtained by transfecting lymphoid cells or injecting with desired Ig genes, obviating the cell fusion process of hybridoma technology.
- A more useful innovation of the technique is the expression of Ig gene in prokaryotic hosts to produce recombinant monoclonals.
- With the advancement in the maintenance of genomic libraries, a combinatorial library can be constructed in *E. coli* to get varieties of antibodies that can be screened.
- A library of innumerable monoclonals can be constructed using M13 phage and the desired antibody can be obtained by screening the library.
- The recent trend is to use transgenic plants to obtain plantibodies or use animals to produce antibodies with desired specificity in milk or hen's egg.

- Rat–rat hybridomas are generally stable and produce high-affinity antibodies.
- Mouse monoclonals can be engineered to suit the human need. Instead of the entire molecule, short fragments containing only Fab or Fv or scFv can be produced with ease.
- Supplementing with the antibody-producing gene through gene therapy is a new move for the treatment of affected persons.

REFERENCES

Barrett Cathy Hefner (1994). "Hybridomas and monoclonal antibodies." In: *Antibody Techniques*. Malik, V.S. and Lillehoj, E.P. (eds.). Academic Press, San Diego.

Goding, J. (1986). *Monoclonal Antibodies-Principles and Practice*, 2nd edn. Academic Press, Inc., London, England.

Harlow, E. and Lane, D. (1988). *Antibodies-A Laboratory Manual*. Cold Spring Harbor Laboratory, Cold Spring Harbor, New York.

Jackson, L. R., Trudel, L. J., Fox, J.G. and Lipman, N. S. (1999). "Monoclonal antibody production in murine ascites-Part I, Clinicopathologic features." *Lab Animal Sci.* **49(1)**: 70–80.

John, McArdle (1998). "Alternatives to ascites production of monoclonal antibodies." *Animal Welfare Information Center Newsletter* **8(3)**: 1–2, 15–18.

Kohler, G. and Milstein, C. (1975). "Continuous cultures of fused cells secreting antibody of predefined specificity." *Nature.* **256**: 495–497.

Kohler, G. and Milstein, C. (1976). "Derivation of specific antibody-producing tissue culture and tumor lines by cell fusion." *Eur. J. Immunol.* **6**: 511–519.

Marx, U., Embleton, M.J., Fischer, R., Gruber, F. P., Hansson, U., Heuer, J., de Leeuw, W. A., Logtenberg, T., Merz, W., Portetelle, D. Romette, J. L. and Straughan, D.W. (1997). "Monoclonal antibody production-The report and recommendations of ECVAM Workshop 23." *ATLA.* **25**: 121–137.

Milstein, C. (1991). "Monoclonal antibodies in immunology-recognition and response." *Reading from Scientific American*. Paul, W.E. (ed.). WH Freeman and Company, New York.

Monoclonal Antibody Production. (1999). A report of the committee on methods of producing monoclonal antibodies-Institute for Laboratory Animal Research National Research Council, National Academy Press, Washington, DC.

Pickup, C. (1985). "Medicine and biotechnology." In: *Biotechnology–Principle and Applications.* Higgins, I.J., Best, D.J. and Jones, J. (eds.). Blackwell Scientific Publications, Oxford, Basel. pp. 83–192.

Wade, N. (1982). "Hybridomas: The making of a revolution." *Science.* Vol. 215: 1073–1075.

3 *IN VIVO* AND *IN VITRO* PRODUCTION OF MONOCLONAL ANTIBODIES

One should consider a number of factors such as concentration, quantity (one or several runs), purity, quality structural characteristics like glycosylation, conformational stability, solubility, specificity and affinity, labour (of maintenance) and cost while selecting the method for large-scale production of monoclonals. The methodology selected must satisfy the needs of its producer and its end user. For less than 10 mg, spent culture supernatant with 'protein A' purification is used. At present, four user groups identified based on the amount of antibody required by them (Figure 3.1 and Table 3.1).

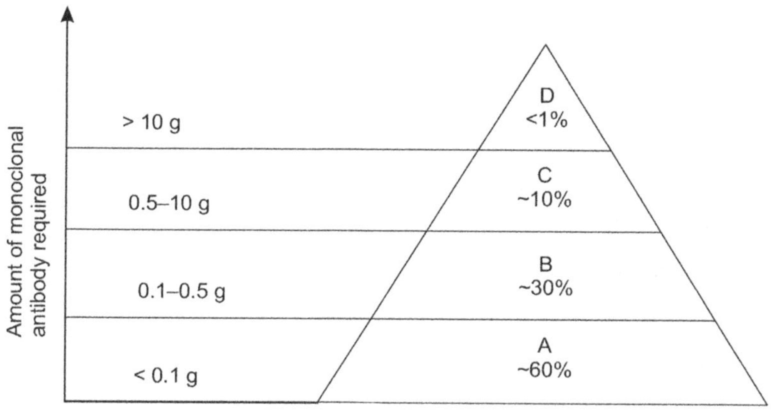

Figure 3.1 Monoclonal user groups (Marx *et al.*, 2006)

Groups A and B adopt the *in vivo* (ascites) technique, since they require monoclonals in small quantities. They use monoclonals for research purposes and not for clinical studies. On the other hand C and D user groups need pure, homogenic and less immunogenic preparations of mononclonals without any contamination of animal viruses. They use monoclonals for preclinical and clinical evaluations and for prophylactic and therapeutic purposes.

Table 3.1 Different categories of monoclonal user groups

Category	% of user group	Method of production	Amount required	Purpose of use of monoclonals
A	60	*In vivo* (ascite) method	Less than 0.1 g	In fundamental and applied researches as analytical and diagnostic kits
B	30	*In vivo* (ascite) method	0.1–0.5 g	In evaluating novel therapeutic monoclonals in animal experiments as diagnostic and analytical kits
C	9	*In vitro* method by large biotech companies	0.5–10 g	Preclinical evaluation of diagnostics and therapeutics in humans
D	< 1	*In vitro* method by large biotech companies	Less than 10 g	Prophylactic and therapeutic purposes

IN VIVO METHOD

Commercial suppliers of monoclonals use the ascites method that has been described briefly in the previous chapter. It is subjected to criticism on both technical and human criteria. The major disadvantages of this animal-based method include the following.

- The animals experience severe cruelty and sufferings as they are routinely subjected to chronic pain and distress. Use of adjuvants further complicates this situation by injuring the animals before the process begins.
- Specific pathogen-free (SPF) animals need to be used and kept quarantined at all times to prevent the contamination of monoclonals. This greatly adds to the cost. Also it is labour-intensive.
- Ascitic fluids may be contaminated with rodent plasma proteins, immunoglobulins (reducing its immunoreactivity) and bioreactive cytokines.
- Extensive animal room facilities, and associated support services are needed with each of the individual animals requiring a daily monitoring (7 days a week).
- Some hybridomas are difficult to grow in rodents.
- Rodents produce monoclonals only for few days.
- 60 to 80 per cent of mice may not produce ascites due to premature death, development of solid tumours or failure to establish *in vivo* hybridoma growth.
- Individual batches of ascites may vary significantly in quality and quantity.

The veterinary problems and pathophysiology associated with the ascites method made The Netherlands Government to introduce a "Code of Practice" in the year 1989, and placed restrictions on its use. The increased humane awareness among Dutch researchers provided further encouragement for the adoption of *in vitro* approaches to monoclonal production, avoiding the use of mice. In 1995, a symposium held in Bilthoven, The Netherlands, concluded that progress in the development of such alternatives (both in efficacy and cost) was sufficient, and the use of ascites was banned first in The Netherlands and later a similar trend was adopted in Germany and Switzerland, followed by Sweden and the United Kingdom.

In October 1996, scientist representatives from several member states of the European Union met at the European Center for the "Validation of Alternative Methods" to discuss the current status of *in vitro* and *in vivo* methods of monoclonal antibody production. After careful consideration of the different types of research and commercial needs for monoclonals and the available production options, they concluded that "for all levels of monoclonal production, there are one or more *in vitro* methods which are not only scientifically acceptable, but are also reasonably and practically available; and as a consequence, *in vivo* production can no longer be justified and should cease." The group further called for a Europe-wide prohibition of the routine use of ascites method for monoclonal production.

In spite of the prohibition of ascites method, it is still in use in many laboratories because of the following reasons:

- more rapid production and high yields of concentrated monoclonals,
- minimal requirements for materials, labour, and technical expertise,
- most hybridomas will grow only in mice,
- relatively less expensive, and
- either an inherent resistance to or lack of familiarity with *in vitro* methods.

In contrast to *in vivo* methods, *in vitro* approaches to monoclonal antibody production have many positive attributes. They are the following.

- The monoclonals derived from *in vitro* alternatives express immunoreactivity in the range of 90 to 95 per cent, regardless of the method used, which is significantly higher than that with the monoclonals produced by ascites.
- *In vitro* cultures rarely fail (3 per cent or less) while a much higher percentage of ascitic mice do not produce antibodies.
- The quality of *in vitro* monoclonals is equal to or better than that derived from the *in vivo* methods.

IN VITRO METHODS

Even though the production of monoclonals is hybridoma-dependent, the choice of equipment, the culture method, type of media, procedural protocols and production parameters are to be carefully selected for *in vitro* methods. It is not always an easy task and depends on the purpose of the production and the product.

Bioreactors for culturing of the animal cells are different from that of the microbes in several respects. Very slow growth rate and product formation, requirement of rich, complex, continuous supply of media, removal of toxic waste products, cells that are sensitive to shearing forces, and rigid asepsis due to easy contamination of cells are the factors to be considered while planning for a large-scale production mode and equipment.

Hybridomas can be cultured as batch, fed-batch, chemostat or perfusion systems. Batch cultures have been still extensively used because of their simplicity and ease of operation. For monoclonal production, fed-batch process has been a good system for most hybridoma cells, since synthesis has been maximum at stationary phase. Unlike batch and fed-batch systems, which are closed systems, a chemostat culture is an open system with an inflow of medium and an outflow of cells and products. This type of system is very effective, since it keeps the cells in a proliferating state. Perfusion systems have been inherently more efficient because they involve savings in media, labour, growth factors and reactor time.

For the production of monoclonals in amounts greater than 100 mg, conventional stirred tank bioreactors of different sizes can be used. In general the maximum antibody concentration achievable is below 100 mg/ml because the concentration of hybridoma cells in suspension cultures hardly exceeds 5×10^6 cells/ml and the supernatant has to be concentrated by ultrafiltration. An increase in cell number is achieved by monitoring and regulating both the pH and dissolved oxygen in the medium. Regulation of pH in high-density cultures is critical to cope with lactic acid accumulation and is regulated with 5% CO_2/air in the bioreactor headspace. Antibody concentration is increased by 2–4 times if the cultures are allowed to grow to exhaustion—over 2–3 weeks without feeding.

According to the principle underlying the culture system, three categories of *in vitro* production system are identified. They are (1) static and agitated suspension cultures, (2) membrane-based and matrix-based culture systems and (3) high cell density bioreactors.

Static and Agitated Suspension Cultures

T-flasks, roller cultures and spinner cultures come under this category. They are easy to handle in cell culture laboratories. They allow a maximum growth of two litres of supernatant per culture unit with a maximum yield of 100–200 mg of monoclonals. Various hybridoma cell lines can be propagated simultaneously. The use of disposable plasticware, serum-free media, low-cost additives like transferrin and insulin, and two low-serum media (a combination of 1% foetal calf serum (FCS) and 0.1% Primatone, a peptic digest of animal tissues), lessens the cost of production.

Tissue culture flasks T-flasks with a maximum working volume of 370 ml are commonly used for monoclonal production. T-flasks can be handled as a static batch culture by placing cells and media in the flask and maintaining in a CO_2 incubator for 7–10 days with little or no monitoring. The flask can be harvested when the media turns yellow (becomes acidic). Monoclonal concentration is typically in the range of 10–100 μg/ml. The initial seeding density, time of proliferation of cell lines and peak antibody level is hybridoma-dependent. T-flasks are frequently used to grow cells for subsequent inoculation into other culture systems.

T-flasks are simple to use, require minimal technical expertise, are relatively inexpensive, are stackable and require minimum incubator space. The principal disadvantage is that the amount of monoclonal is very low, necessitating concentration.

Roller bottle cultures and spinner flasks The cultures are agitated in roller bottles and spinner flasks. As a result, the cells grow to high cell density in suspension cultures than in stationary cultures of T-flasks. Monoclonal concentrations obtained with these two methods are typically greater, ranging from approximately 10–220 μg/ml, and the average duration of the culture is longer, approximately 12 days. Culture volumes are typically $\leq$ 2 L. Advantages and disadvantages are similar to those described for T-flasks, but roller bottles and spinner flasks require more incubator space, and are more expensive than T-flasks.

Gas-permeable cell culture bags Flexible, gas-permeable, pre-sterilized, disposable cell-culture bags with attached ports and tubing with roller clamps (Figure 3.2) that are used for inoculation of cells and media, for sampling during production and for harvesting, have been commercially available for monoclonal production.

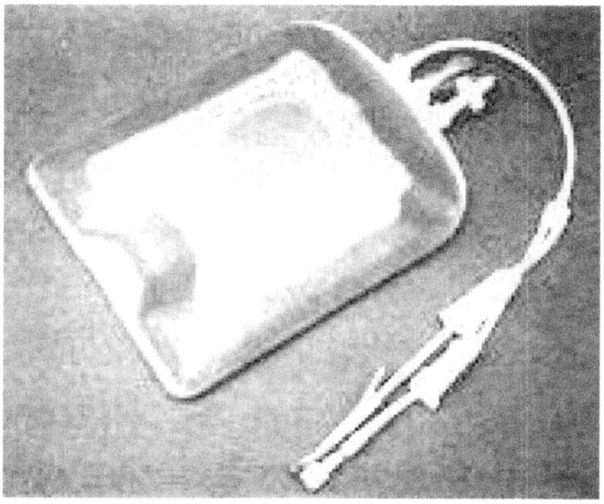

Figure 3.2 Gas-permeable cell culture bag

Cell culture bags have the following advantages over flasks.

- Bags have more surface area for CO_2 and oxygen diffusion, improving cell oxygenation and viability.
- They are completely enclosed systems, thus microbial contamination is reduced.
- They may be placed flat on the incubator floor or hung on a stand, potentially reducing the required incubator space.

Wave bioreactor The flexible, plastic pre-sterilized, disposable wave bioreactor cell culture bag is provided with an inlet air filter, exhaust air filter, a needle-less syringe fitting for sampling from the bag, a tubing connector for additions to, and harvests from, the bag. It is partially filled with the culture media and cells, inflated with gas to rigidity and maintained on a rocking platform in a CO_2 incubator (Figure 3.3). The headspace in the bag is continuously aerated. The agitation generated by the rocking mechanism causes wave action that increases oxygen transfer and mixing in the bioreactor and suspends the cells with low shear. A laminar flow hood has not been required for adding to, or sampling from, the bag.

Figure 3.3 Wave bioreactor (Jackson *et al.*, 1999)

Two-litre bags (0.1–1-L culture volume) and 20-L bags (1–10-L culture volume) have been suitable for small-scale monoclonal production. The wave bioreactor rocking unit with holder can accommodate two 2-L bags or one 20-L bag with adjustable rocking speed and with a digital read-out. An electrical air pump keeps the bag aerated. A bench-top model with a heater and controller, as well as a CO_2/air mixing unit, has also been available. The bioreactor may be handled as a batch or fed-batch culture system. In the fed-batch mode, the volume of media may be increased gradually as the cell density increases.

Membrane-Based and Matrix-Based Culture Systems

For user groups A, B and C (who require up to 10 g of monoclonals), membrane-based and matrix-based static cultures as well as suspension bioreactors can be used.

In the membrane-based systems, the cells are cultured in the compartments separated from the nutrient supply by perfusion membranes or special gassing membranes that enhance the oxygen transfer. They are easy to handle and enable various cultures to be run simultaneously in routinely equipped cell culture laboratories.

In matrix-based systems such as fluidized bed or ceramic bioreactors, the immobilization of cells on matrices enables them to be perfused actively and continuously with fresh medium. In most cases, the supernatant produced has to be concentrated by precipitation or ultrafiltration before carrying out the special purification procedures. Special training is required for the proper handling of these systems.

CELLine culture systems In the culture devices made of polystyrene and pre-sterilized disposable materials, the cells and secreted monoclonals are retained in a small-volume cell compartment, separated by a semipermeable membrane from the larger volume of the nutrient medium compartment above and a gas-permeable membrane below. The nutrient medium compartment has a wide mouth port with a screw cap through which medium can be added and a cell compartment port with a screw cap and septum through which cells can be added or drawn via pipette (Figure 3.4).

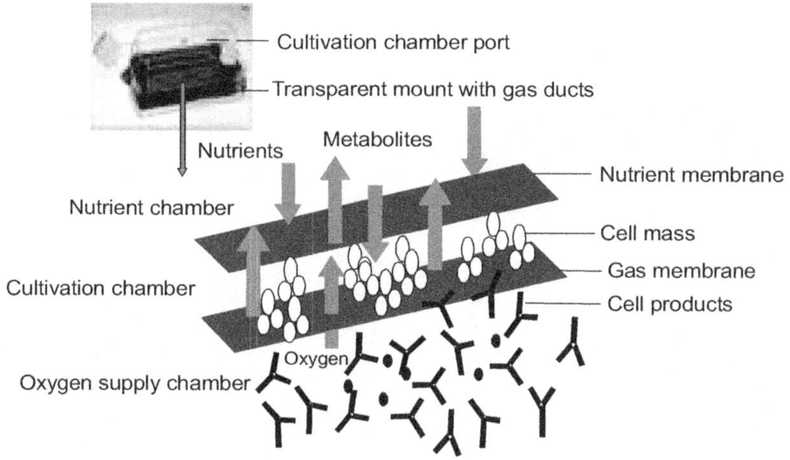

Figure 3.4 The CELLine culture system (Jackson, *et al.*, 1999)

The smaller volume of cell compartment provides higher densities of cells and significantly higher concentrations of monoclonals, as compared to stationary culture techniques, minimizing downstream concentration steps. The systems can be handled easily and used as interlocking units and stackable to CO_2 incubator with less space. Since serum can be supplemented in the small-volume cell compartment, the cost of production is reduced considerably.

Several sizes and types of cell line culture devices can be used for different purposes. The CL 6 has six parallel chambers, each with a 0.75-ml cell compartment and a 5–30-ml nutrient medium compartment, useful for molecular biology studies, for screening different clones, for optimizing medium and for pharmacological testing. The CL 350 has a 5-ml cell compartment and a 50–350-ml nutrient media compartment and is useful for small-scale monoclonal production.

The CL 1000 has a 15-ml cell compartment and a 100–1000 ml nutrient medium compartment for large-scale monoclonal production.

Mini PERM bioreactor The mini PERM bioreactor has been a modular mini fed-batch fermenter. This system has a disposable 40-ml production module containing hybridoma cells and a nutrient reservoir of 550-ml capacity, separated by a semipermeable dialysis membrane. The dialysis membrane allows exchange of nutrients, metabolic wastes, and dissolved gases. A gas-permeable silicon membrane with baffles in the production module permits gas exchange. Luer-lock connections on the production module have been used for cell inoculation and sample collection.

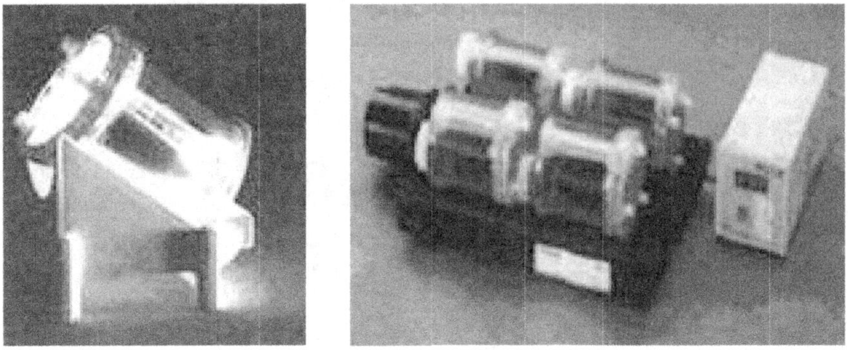

Figure 3.5 Mini PERM bioreactor and four mini PERM bioreactors on a bottle-turning device with remote control unit (Jackson, *et al.*, 1999)

To agitate the nutrient medium and to facilitate the passage of nutrients and metabolites across the dialysis membrane, the mini PERM bioreactor (Figure 3.5) is designed to rotate on a bottle-turning device within a CO_2 incubator. A larger bottle-turning device that accommodates up to four bioreactors is also available. The production module is pre-sterilized and disposable, and the nutrient reservoirs can be autoclaved and reused up to 10 times.

Reported advantages of this system include

- growth of cells at high cell densities ($\geq 10^7$ cells/ml),
- high monoclonal concentration and high product purity related to serum reduction or removal from the nutrient compartment,

- relative ease of use,
- ability to maintain the cultures for a relatively long period of time and
- ability to reuse some components.

High Cell Density Bioreactors

In these culture systems, cell densities greater than 10^8 cells/ml have been achieved. User groups B and C have been adopting this system due to the high antibody concentration.

Hollow fibre bioreactors In the hollow fibre bioreactor, the culture medium is passed through bundles of hollow fibres—kind of *in vivo* capillary systems—providing a more stable pericellular microenvironment for cultured cells with regard to nutrient supply, metabolic waste removal and pH. The basic design of a bioreactor circuit is illustrated schematically in Figure 3.6.

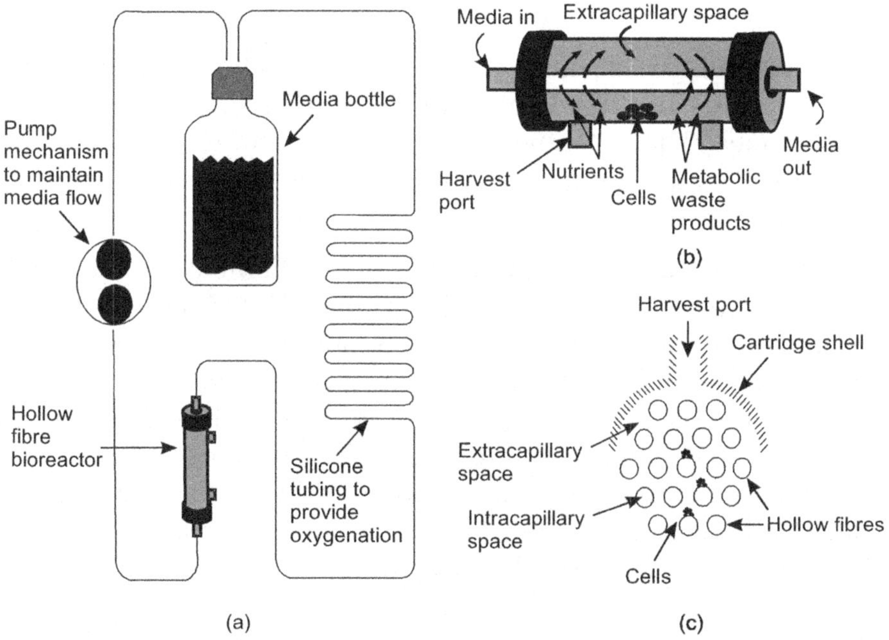

Figure 3.6 Hollow fibre bioreactor (a) Longitudinal view (b) Circular view (c) Cross-sectional view (Jackson, *et al.*, 1999)

A pre-sterilized and disposable media bottle with a sampling port, a variable-speed pump to maintain continuous and unidirectional media flow, an HFB cartridge and a means of providing oxygen and CO_2 exchange, is included in this system. The culture media is continuously perfused in the intracapillary space and the cells are grown in the extracapillary space. The walls of the hollow fibres serve as semipermeable ultrafiltration membranes that retain the cells and secrete monoclonals in the relatively small volume of the extracapillary space, while permitting gas, nutrients, and metabolic waste products to diffuse freely across the membrane due to hydrostatic pressure differences and concentration gradients. The extracapillary space has harvest ports for examination of cells and monoclonal harvest.

The advantages of HFB systems are the following.

- Cells are protected from shear.
- Cells can be grown to very high densities (10^7–10^8 cells/ml).
- Cell viability and production can be maintained for extended periods of time, ranging from weeks to months.

The highly concentrated monoclonal product in a small volume necessitates the need for extensive downstream concentrating techniques. The potential for mechanical failure, and the need for technical expertise and familiarity with the system are the few disadvantages.

Cellmax artificial capillary cell culture system Cellmax artificial capillary cell culture system includes either a single HFB module with a pump station and power supply. Four HFB modules on the base unit, each with independent media bottle and flow-path, with a central pump station are also available (Figure 3.7). The HFB modules are easily disconnected from the pump station for transport to a laminar flow hood for manipulations. An AC motor cable connects the base unit to an electronic control unit that is placed outside the incubator and connected to a standard electrical outlet.

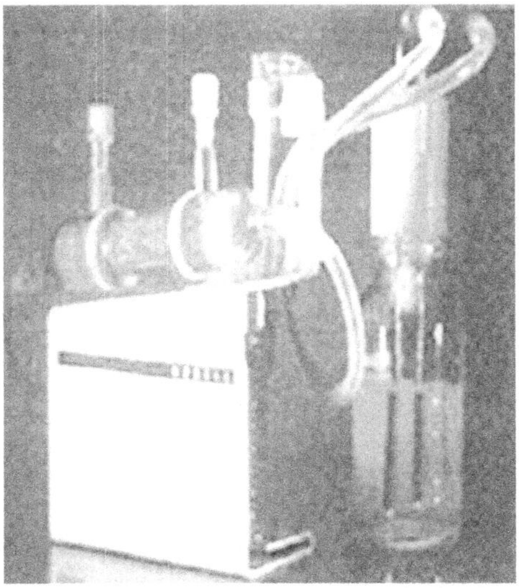

Figure 3.7 Cellmax artificial capillary cell culture system

Cell-Pharm system CP 100 It is a bench-top culture system with a media-heating block, as well as temperature and CO_2 control, eliminating the need for a CO_2 incubator. The HFB with media flow-path and oxygenator, clip easily on to the unit. Bottled or bagged media may be used. The monoclonal product can be harvested manually via positive displacement through the harvest port septum or via a Cell-Pharm autoharvester (Figure 3.8). A laminar flow hood is not required for manipulation of the system.

Figure 3.8 Cell-Pharm system CP 100 with autoharvester (Jackson, *et al.*, 1999)

TECNOMOUSE It has an instrument module with a removable rack for placement of one to five bioreactors, each with independent media flow-path, and a pump station which contains individual pump cassettes for each bioreactor (Figure 3.9). The fibres of HFB are arranged in single parallel rows and are encased in a silicone membrane, within a flat rectangular plastic cartridge. Direct gassing to the culture is provided by individual gassing port, where incubator air is pumped to channels outside the gas-permeable silicone membrane.

The module can be programmed to change the media flow rate in fixed or variable increments over a specified time interval. A laminar flow hood and CO_2 incubator are not required for manipulation of the system, as a thermo-hood has been available for bench-top operation.

Figure 3.9 The TECNOMOUSE system (Jackson, *et al.*, 1999)

Other *in vitro* techniques include growing cells in dialysis tubing within a culture bottle, and use of oscillating bubble dialysis chambers or tumbling chambers. Laboratory-scale stirred tank reactors, fermenters, ceramic-matrix bioreactors and packed-bed bioreactors are also in use.

Microencapsulation technique This is an interesting method patented by the Damon Corporation in which hybridoma cells are grown inside hollow microspheres which are surrounded by a porous membrane. Being protected inside the microsphere, the cells multiply and grow to higher cell densities. Since the antibody is retained by the membrane, it does not become contaminated with other immunoglobulins if serum has to be included in the growth medium. Using this (Figure 3.10) microencapsulation technology, 100 mg to 1 g of antibody can be produced per litre of culture medium.

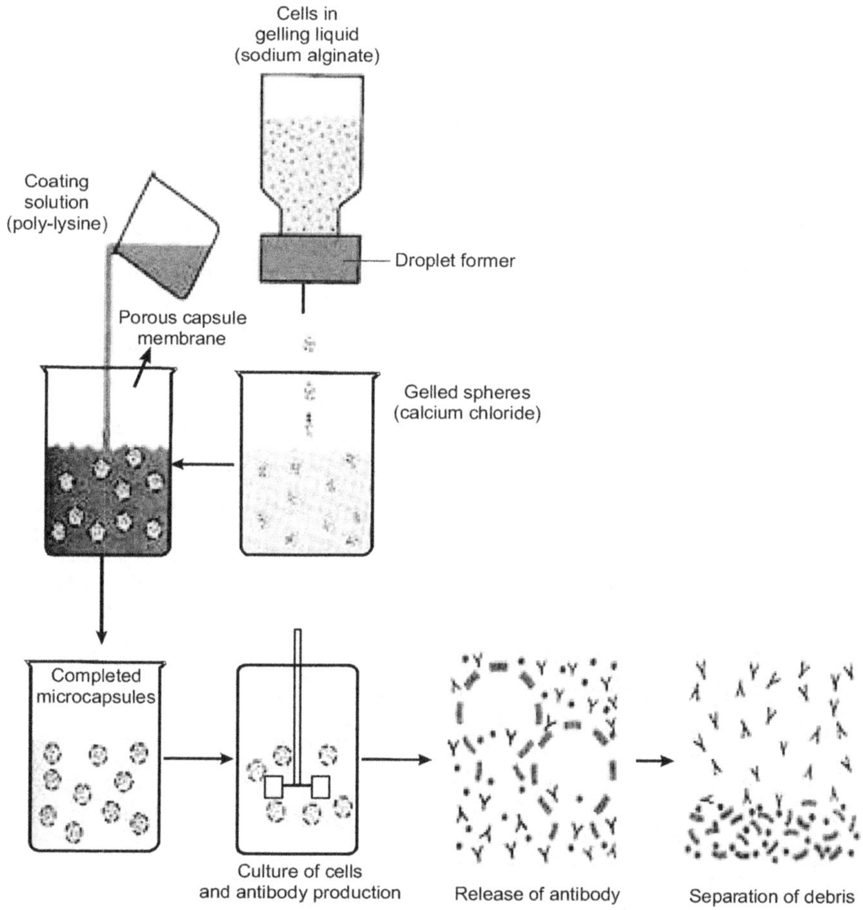

Figure 3.10 Microencapsulation method where hybridomas grow in microspheres (Primrose, S.B., 1999)

LARGE-SCALE PRODUCTION

To date, industrial production has been involving fermenter-based systems that show good scale-up potential. Scale-up can be achieved either as a multiple-process system or as a unit-process system. The multiple-process system is defined as one that is scaled up by increasing the number of units. Such a procedure has a number of disadvantages, including higher capital and operating costs, less controllability, and poorer operating conditions.

In contrast to the multiple-process system, the scale-up of the unit-process system is achieved by increasing the size of the culture equipment without a substantial increase in the number of vessels. In the unit-process system, process parameters such as temperature, pH, dissolved oxygen, glucose, lactate and certain amino acids can be monitored.

RESOURCES OF INFORMATION

Two useful resources for finding and purchasing commercially available monoclonals are the "Antibody Resource" website, which contains links to many commercial monoclonal suppliers and "Linscott's Directory of Immunological and Biological Reagents".

The International Hybridoma Data Bank (IHDB) provides a comprehensive directory of information on hybridomas and other cloned cell lines and their immunoreactive products, such as monoclonals. The American Type Culture Collection (ATCC) is responsible for collection and dissemination of the data contained in this resource. Information on hybridoma construction, the reactivity and non-reactivity of the secreted monoclonals and the availability of individual hybridomas and their monoclonal products are included in the database.

REFERENCES

Anon. (1989). *Code of practice for the production of monoclonal antibodies*. Veterinary Health Inspectorate 6. Rijswijk. The Netherlands.

Brodeur, B. and Tsang, P. (1986)."High yield monoclonal antibody production in ascites." *J. Immunol. Methods*. **86**: 239–241.

Chandler, J. (1987) "Factors influencing monoclonal antibody production in mouse ascites fluid." In: *Commercial Production of Monoclonal Antibodies*. Seaver, S.(ed.). Marcel Dekker, New York. 75–92.

de Geus, B. and Hendriksen, C. (1998) "*In vivo* and *in vitro* production of monoclonal antibodies–Introduction." *Res. Immunol.* **149:** 533–534.

Evans, T. and Miller, R. (1988). "Large-scale production of murine monoclonal antibodies using hollow fibre bioreactors." *Biotechniques.* **6**: 762–767.

Falkenberg, F., Weichert, H. and Krane, M. (1995). "*In vitro* production of monoclonal antibodies in high concentration in a new and easy to handle modular minifermentor." *J. Immunol.* **179**: 13–29.

FDA (Food and Drug Administration, Center for Biologics Evaluation and Research). (1997). Points to consider in the manufacture and testing of monoclonal antibody products for human use, Washington, DC.

Federspiel, G., McCullough, K.C. and Kihm, U. (1991). "Hybridoma antibody production *in vitro* in type II serum-free medium using Nutridoma-SP supplements. Comparison with *in vivo* methods." *J. Immunol.* **145**: 213–221.

Heidel, J. (1997). Monoclonal antibody production in gas-permeable tissue culture bags using serum-free media. Proceeding for Alternatives in Monoclonal antibody production–Center for Alternatives to Animal Testing. **8**: 18–20.

Hendriksen, C., Rozing, J., Vander Kamp, M. and deLeeuw, W. (1996). "The production of monoclonal antibodies: Are animals still needed?" *ATLA.* **24**: 109–110.

Hendriksen, C. and de Leeuw,W. (1998). "Production of monoclonal antibodies by the ascites method in laboratory animals." *Res. Immunol.* **149**: 535–542.

Jackson, L.R., Trudel, L.J., Fox, J.G. and Lipman, N.S. (1996). "Evaluation of hollow fiber bioreactors as an alternative to murine ascites production for small scale monoclonal antibody production." *J. Immunol. Methods.* **189**: 217–231.

Jackson, L.R., Trudel, L.J. and Lipman, N.S. (1999). "Proceedings of the production of monoclonal antibodies workshop-small-scale monoclonal antibody production. *In vitro:* methods and resources." *Lab Anim. Sci.* **28**: 3.

Jackson, L.R., Trudel, L.J., Fox, J.G. and Lipman, N.S. (1999). "Monoclonal antibody production in murine ascites: Part II production characteristics." *Lab. Animal Sci.* **49**(1): 81–86.

Kamp, M. and de Leeuw, W. (1996). "Short review of *in vitro* production methods for monoclonal antibodies." *NCA Newsletter.* **3**: 10–11.

Klerx, J. Jansen, Verplanke, C., Blonk, C. and Twaalfhoven, L. (1988). "*In vitro* production of monoclonal antibodies under serum-free conditions using a compact and inexpensive hollow fibre cell culture unit." *J. Immunol. Methods.* **111**: 179–188.

Kuhlman, I., Kurth, W. and Ruhdel, I. (1989). "Monoclonal antibodies: *in vivo* and *in vitro* production on a laboratory scale, with consideration of the legal aspects of animal protection." *ATLA.* **17**: 73–82.

Lipman, N. (1997). "Hollow fibre bioreactors: An alternative to the use of mice for monoclonal antibody production." In: *Alternatives in Monoclonal Antibody Production*. Johns Hopkins Center for Alternatives to Animal Testing Technical Report. **8**: 10–15.

Marx, U., Embleton, M.J., Fischer, R., Gruber F.P., Hansson, U., Heuer, J., de Leeuw, W.A., Logtenberg, T., Merz, W., Portetelle D., Romette, J.L. and Straughan, D.W. (2006). Monoclonal antibody production-the report and recommendations of EVAM Workshop, 23.

NIH (National Institutes of Health). (1997). "Production of monoclonal antibodies using mouse ascites method." 98–01. Rockville, MD: Office for protection from research risks, Nov. 17.

Peterson, N. and Peavey, J. (1998). "Comparison of *in vitro* monoclonal antibody production methods with an *in vivo* ascites production technique." *Contemporary Topics Lab. Anim. Sci.* **37**(5): 61–66.

Primrose, S.B. (1999). *Modern Biotechnology.* Blackwell Scientific Publications, Oxford, London.

Tarleton, R. and Beyer, A. (1991). Medium-scale production and purification of monoclonal antibodies in protein-free medium." *Biotechniques.* **11**: 590–593.

4 GENETIC ENGINEERING OF MONOCLONALS

Monoclonals produced by hybridoma technology are usually mouse proteins. While mouse monoclonals have revolutionized biology in general and immunology in particular, there are some limitations when it comes to using them for injection into humans. Monoclonal antibodies of mouse and human are not identical since there are numerous differences in their amino acid sequences. These sequences account for the immunogenicity. The mouse antibody of potential therapeutic value when injected into a human will eventually be recognized as a foreign protein, will elicit immune response by producing human anti-mouse antibody or HAMA response and will soon be cleared from the circulation.

The production of human monoclonal antibodies by conventional hybridoma technique has the following problems.

- For ethical reasons, it is not acceptable to immunize human beings.
- It is difficult to obtain specific human B lymphocytes from the peripheral blood. Also, a majority of them express surface IgM that interferes with the affinity of monoclonals.
- No effective human myeloma cell lines that can replace the mouse myeloma cell lines have been discovered. Human myeloma cell lines have poor growth characteristics and usually produce their own antibody which complicates production and purification.
- Human hybridomas/heterohybridoma (human–mouse) cell lines are highly unstable and so cells producing a human monoclonal are extremely rare.
- It is difficult to select high-affinity monoclonals with specific isotype.

Therefore, the sheer explosion and adoption of this technology in almost every immunology lab led to the development of alternative technologies for generation of human monoclonals, following the original method. These alternative technologies have been discussed briefly in chapter 1 and are dealt with elaborately in this chapter.

EBV TRANSFORMATION

The Epstein–Barr virus is used to transform human B cells to generate human monoclonal antibodies. Human B lymphocytes that actively produce antibodies are isolated from the subjects and are treated with a fluorescence-labelled antigen. Fluorescent-activated cell sorting is then used to enrich the cell sample for specific antibody-producing B lymphocytes. They are then transformed with Epstein–Barr virus, which allows them to grow readily in culture. Some of these transformed B cell clones secrete human monoclonal antibodies that interact with the selected antigen. Unfortunately, with this strategy, the yields tend to be small and the monoclonals have weak antibody-binding affinities. In addition, there is a low probability that a non-immunized individual may have antibody-secreting cells that recognize the selected antigen.

ONCOGENIC TRANSFECTION

As a variation of the EBV transformation technique, oncogenic DNA of lymphocytes is employed to generate continuous or immortal cultures of antibody-secreting transfected cells with varying degrees of success.

HUMAN–HUMAN HYBRID

Attempts have been made to fuse sensitized human B cells with human plasmacytoma or with lymphoblastoid cell lines. But to obtain a non-antibody-secreting human lymphoblastoid cell is difficult.

INTERSPECIFIC HYBRID

Fusing the human B cells with non-secreting mouse or rat myeloma cells, one can get interspecific myeloma cells. The B cells secreting specific antibodies are quite few in peripheral blood, but their number can be increased by stimulating the growth of these cells *in vitro* with pokeweed nitrogen or an antigen or a combination of both.

FROM SEVERE COMBINED IMMUNODEFICIENCY (SCID) MOUSE

It may also be possible to introduce cells of the human immune system into a mutant mouse strain that lacks for the most part its own natural immunological cell repertoire. Such a mouse is called SCID. After the transplantation of stem cells of the human immune system, it can produce human antibodies in response to the challenge by antigen.

MONOCLONALS FROM RECOMBINANT DNA TECHNOLOGY

A more recent innovation has been the technique of expressing transfected Ig gene into prokaryotic or eukaryotic hosts to produce genetically engineered monoclonals. This technique bypasses the fusion step which is inefficient and laborious in hybridoma production. Such genetically engineered cells are called **transfectomas**.

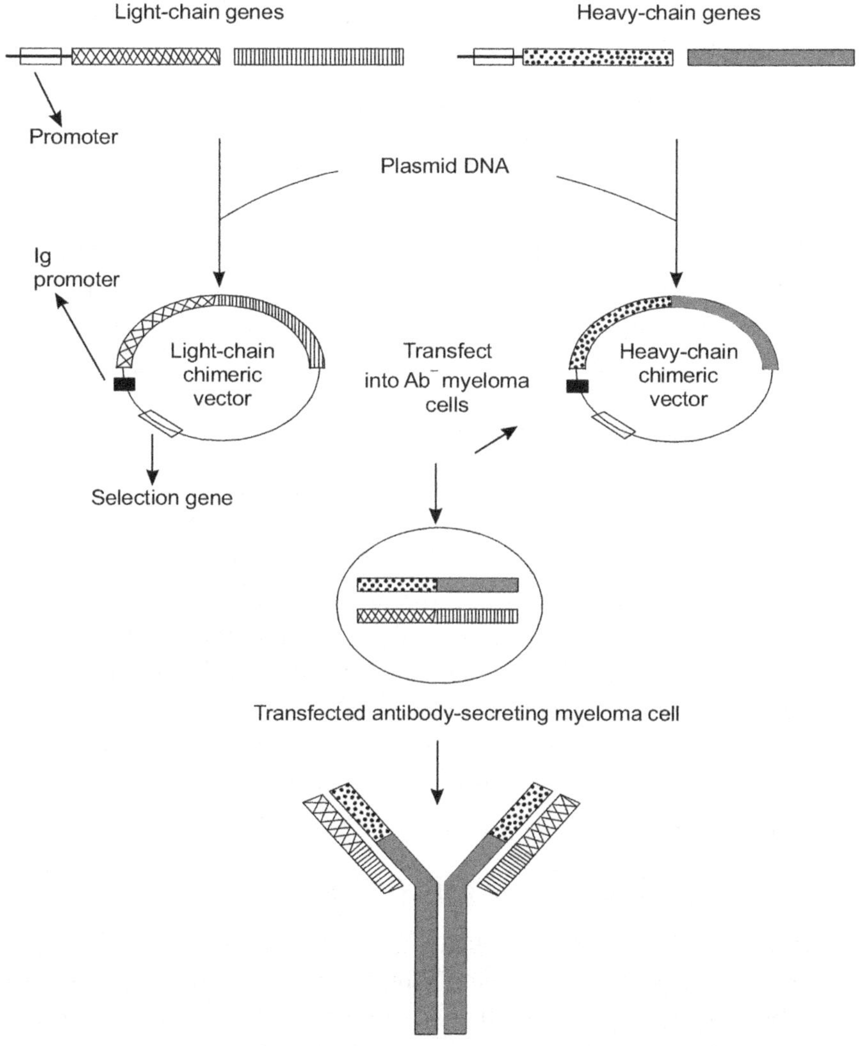

Figure 4.1 Production of chimeric mouse–human monoclonals (Kuby, J., 1997)

Engineering of Chimeric Mouse–Human Antibody

To reduce the immunogenicity of mouse monoclonal antibodies, chimeric genes are used. The cDNA fragment encoding the mouse V_L domain with a promoter and leader, is ligated to the fragment encoding the human C_L domain. Similarly mouse V_H genes are ligated to human C_H genes. Because the C domain does not contribute to antigen-binding, the chimeric antibody obtained by this combination will retain the same antigen-binding specificity as the original mouse monoclonal, but will be closer to human antibody in sequence.

The chimeric mouse–human genes are introduced into two separate expression vectors. These vectors are transfected into Ab^- (non-antibody-secreting) myeloma cells and cultured in the presence of ampicillin medium. Ampicillin selects the transfected myeloma cells that secrete the chimeric antibody (Figure 4.1). This chimeric antibody is still not fully humanized because it may still contain the idiotope (epitope), the amino acid sequence from the mouse protein of antibody near the antigen-binding site.

Fully Humanized Antibody

The fully humanized antibody is obtained by transferring the complementarity-determining regions (CDRs) of the rodent monoclonals to a natural human antibody. To make a fully humanized antibody, *in vitro* (oligonucleotide-directed) mutagenesis is used. In six PCR cycles, the human CDR is replaced by the mouse CDR. In other words, six CDRs of mouse are grafted in the **framework region** (FWR) of human variable region of H and L chains. These reshaped human antibodies have antigen-binding affinities similar to those of rodent monoclonal antibodies. They may be used as effective therapeutic agents as there no immunogenic reaction.

COMBINATORIAL LIBRARY IN *E. COLI*

A mouse is inoculated with an antigen. Spleen cells are recovered from the immunized mouse. The mRNA is obtained and cDNA is synthesized with reverse transcriptase. The heavy- and light-chain genes are separately amplified by PCR, and cDNA is ligated into λ cloning vectors. Two different libraries are produced, one containing H-chain genes and another containing L-chain genes. (This step has been omitted from the figure for simplicity.)

Phage DNA is isolated from each library and the H- and L-chain sequences are ligated together and packaged to form a combinatorial library. Each phage now contains a random pair of cDNAs of H and L chains and thus upon infection of *E. coli* produces two antibody chains. Since the heavy chain sequences contain only the variable region and the first constant domain, the antibody that forms is called Fab (Figure 4.2). It binds to an antigen much like an intact antibody but lacks the effector domains.

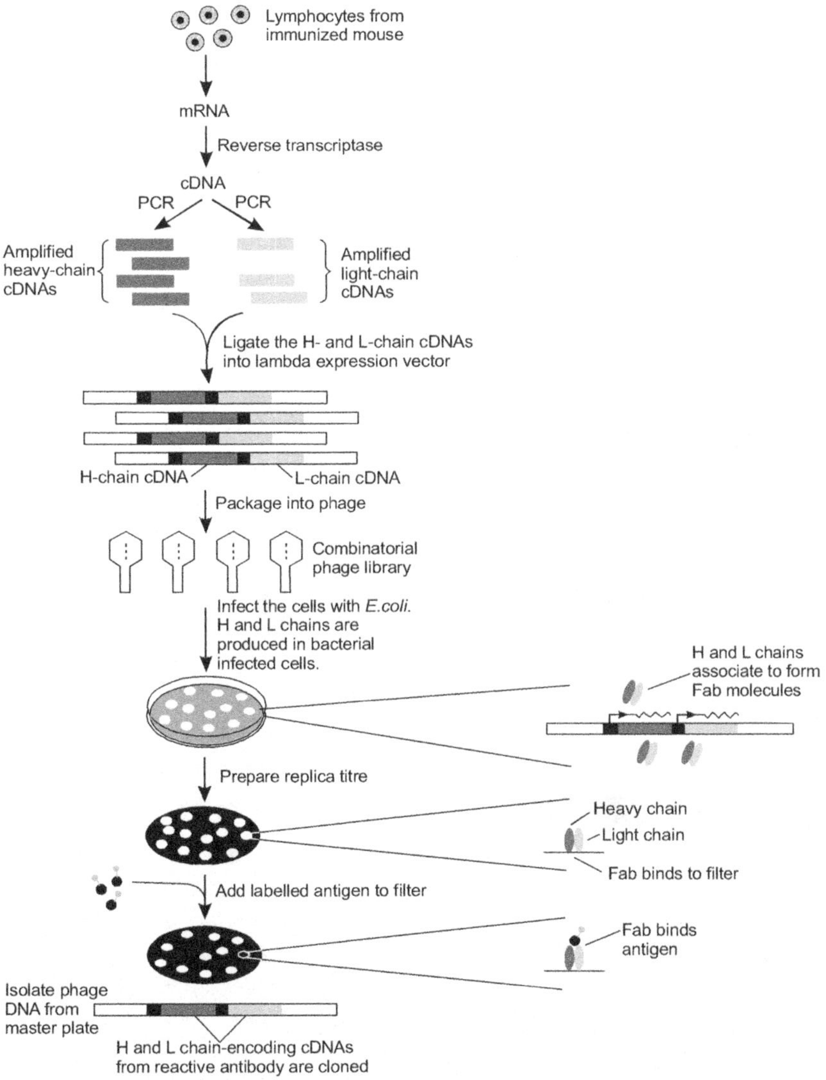

Figure 4.2 Combinatorial library of antibodies expressed in *E.coli* (Watson, *et al.*, 1992)

The combinatorial library is plated on a bacterial lawn, and the resulting phage plaques, each containing a unique antibody, are screened with radiolabelled antigen. Out of million phage plaques screened, 200 clones were identified that produced the antibody binding the antigen. Thus with this approach, million antibodies are possible (at least 100 times more than from conventional monoclonal antibody technique).

PHAGE DISPLAY

A recent modification of this combinatorial library uses filamentous phages such as M13 instead of λ phage and allows display of the antibodies on the phage surface (Figure 4.3). This offers the advantage of being able to screen more than thousand phages because the screening can be done in solution, and of selecting the phage that can express protein-bound antibodies.

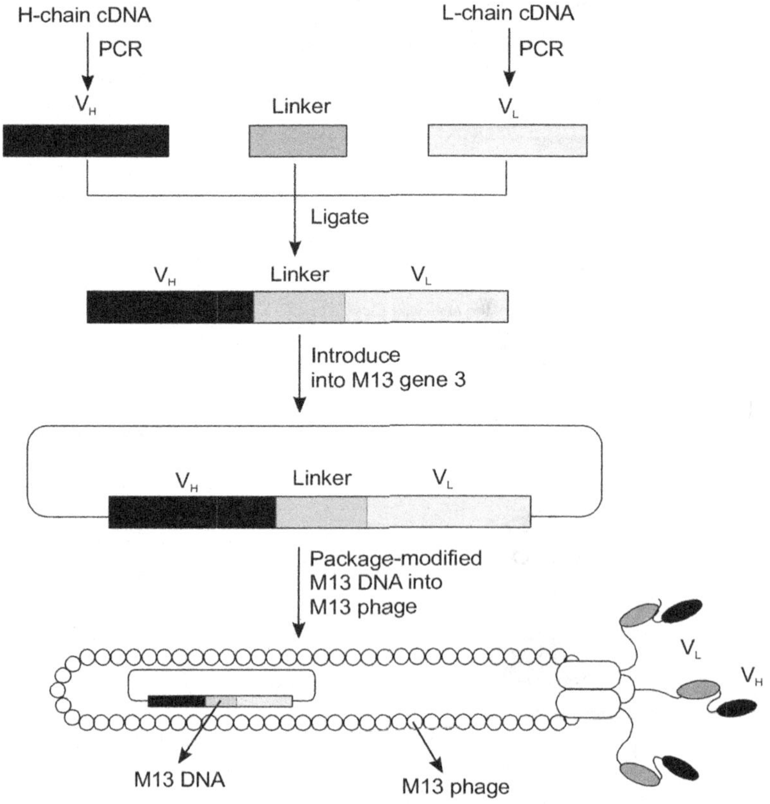

Figure 4.3 Construction of phage display library

The cDNAs for the V_L and V_H regions are amplified by PCR and then ligated to the DNA along with a short linker peptide. The DNAs encoding the single chain fragment variable (scFv) or Fab combinatorial library are introduced into the genome of M13 and fused to gene 3 of phage which encodes a phage surface protein. The gene 3 expresses three surface proteins on each phage and each protein expresses along with the reombinant ScFv or Fab. Hence each phage produces three combinatorial ScFv or Fab.

The phage with the appropriate antibody fragments displaced is then selected by binding to the antigen attached to a solid support. Usually several rounds of selection are carried out to isolate high-affinity antibody fragments. The selected phage can then be used to express soluble antibody fragments and to recover the antibody genes which are used to construct the desired form of an antibody.

The phage display approach offers several potential advantages over hybridoma technology. New antibody specificities can be isolated rapidly, with or without prior immunization, and high-affinity antibodies are obtained. Generation of human antibodies even to auto antigens is possible, which is difficult to achieve by other means. Antibodies to weak epitopes can be deliberately selected by masking immunologically dominant epitopes.

In addition it is possible to use antibody chains isolated by hybridoma techniques in chain shuffling or mutagenesis experiments with phage libraries and to isolate novel high-affinity antibody chains.

ENZYMATIC CLEAVAGE OF ANTIBODIES

Antibody molecules are very amenable to manipulation enzymatically. They can be cleaved into their constituent fragments by immobilized pepsin or papain. These fragments, either alone or pegylated with some other substances, appear to have significant advantages in radio-immunoimaging because their reduced size allows a better tissue penetration, shorter blood residence time and a reduction in the HAMA response.

Fragments Obtained from a Monoclonal Antibody

Mild chemical reductions of some antibodies have been known to result in the separation of the two heavy chains to form half molecules with univalent binding activity. Another type of monovalent antibody fragment termed Fab/c, is prepared from rabbit antibodies by partial digestion with papain to cleave only one Fab arm leaving a molecule with Fc and Fab as a structural and functional unit (Figure 4.4). This type of molecule like the monovalent monospecific monoclonals was shown to exhibit an enhanced cell killing both *in vitro* and *in vivo* perhaps by avoiding antigenic modulations.

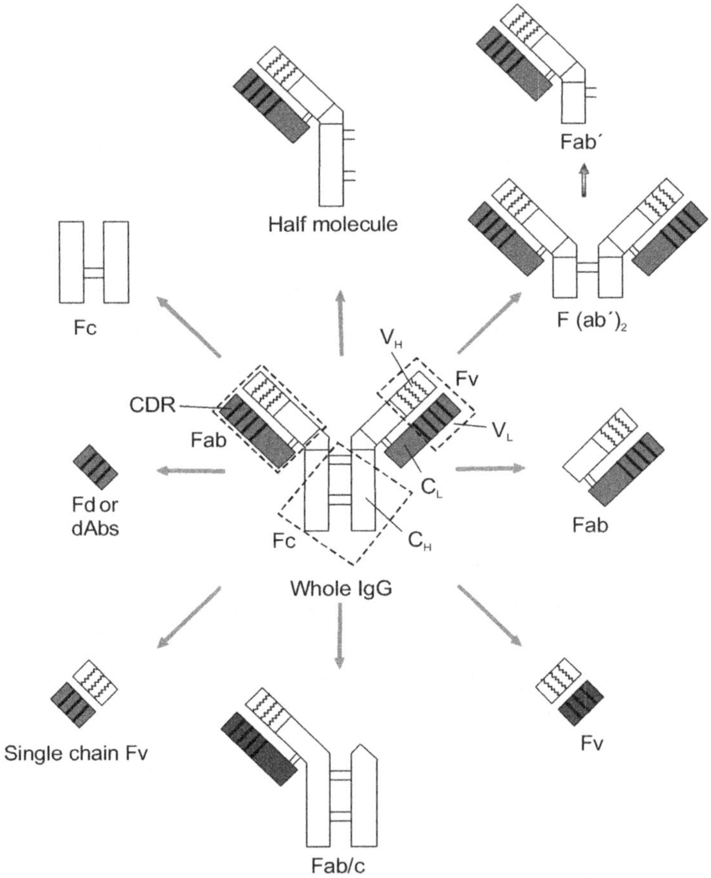

Figure 4.4 Types of fragments obtained from a monoclonal antibody (Suresh *et al.*, 1991)

The amino terminal half of the Fab region representing the variable domains of the two chains is referred to as Fv and such fragments are prepared by enzymatic cleavage and more recently by expressing the V_H and V_L gene segments in *E. coli*. A further nuance of generating such binding domains was reported in that V_H and V_L gene segments are joined together with intervening sequences. The expression of such a construct resulted in the synthesis of single chain Fv (scFv), as one polypeptide chain folding back on itself to form an active binding domain. The recombinant DNA approach is also utilized to clone the V_H region or single domain antibodies (dAbs) from the splenic DNA of mice.

The two chains of an Fv fragment are less stably associated than the Fd and Fab fragments. To stabilize an Fv fragment, two strategies are adopted. First, by mutating a particular residue to cysteine for the formation of disulphide bonds between the two domains, disulphide-linked Fv is obtained. The second is the introduction of a peptide linker between the C-terminus of one domain and the N-terminus of the other, such that the Fv is produced as a single polypeptide chain known as scFv (Figure 4.5). The production of scFv molecules requires a suitable peptide linker to span a 35–40 Å distance allowing a correct folding and assembly of the Fv structure.

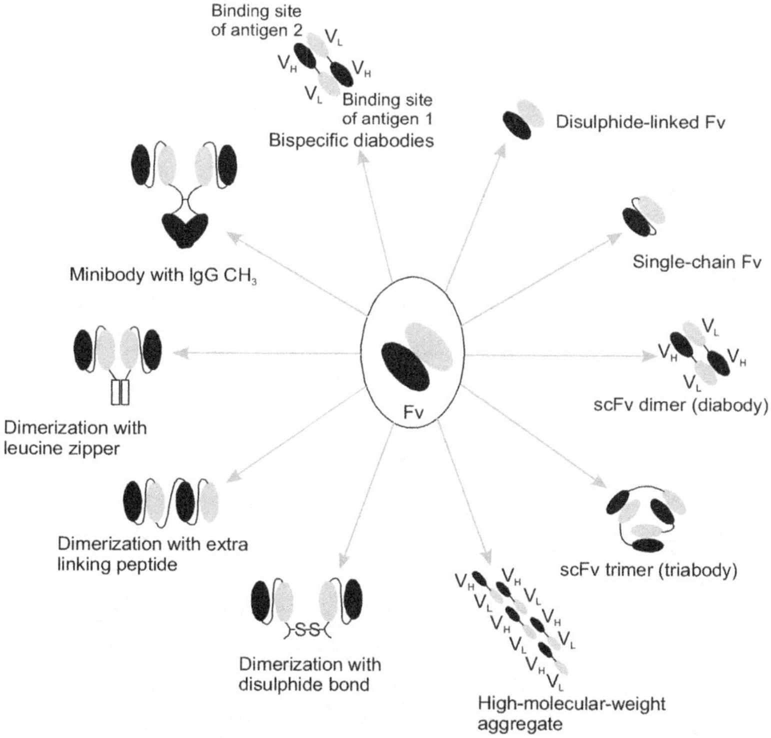

Figure 4.5 Strategies used for the development of different kinds of Fv. (modified from King, D.J., 1998)

As the chain length of the linker is decreased, there is a tendency towards dimer formation, with very short linker or no linker at all. This leads to the production of stable dimeric structures termed **diabodies** (Figure 4.5). A diabody is functionally better than scFv. Its bivalency increases the avidity. Diabodies with two different affinities can be made as 'bispecific'. In some cases, particularly with direct fusion of V_H and V_L, stable trimeric species are produced resulting in **triabodies**. Some triabodies show improved avidity. Many scFvs are shown to be susceptible to aggregation with dimers, resulting in **high-molecular-weight aggregates**.

Several alternative approaches have been adopted in attempts to produce dimeric species of scFv. They are produced by

i. forming a disulphide bond or by chemical cross-linking through attached hinge-region peptide.
ii. fusing one scFv to another directly through an extra linking peptide.
iii. attaching scFvs to dimerizing domains like leucine zippers and
iii. producing minibody by linking two scFvs by CH_3 domains and a hinge region peptide (Figure 4.6).

Bispecific diabodies are the diabodies with two different antigenic specificities. They are formed by the association by two different ScFv fragments.

Trimeric and tetrameric antigen-binding antibodies can also be produced by chemical cross-linking of Fab or scFv with tri- or tetra-maleimide cross-linking reagents. Tri-scFv or tri-Fab produced in this way show increased avidity of binding to antigen.

ANTIBODY (PROTEIN) ENGINEERING

Antibody molecules are made up of a series of domains which are capable of folding into their native structure independently. This has enabled many manipulations of the antibody molecule to be carried out without affecting the confirmation of adjacent domains, and allows domains to be moved in position, or substituted with other molecules very easily. The beta-pleated sheet structure allows substitution of the CDR loops. It is possible to switch one antibody from another without disturbing the FWR. This ease of genetic manipulation has led to the technology of antibody engineering that allows the specific design for specific applications.

The specificity and affinity of the binding site, valency and size, requirement of effector function, attachment of effector or reporter molecules, immunogenicity and cost of production are several factors that must be considered while designing an antibody. The types of engineered monoclonals shown in Figure 4.6 are discussed in the following section.

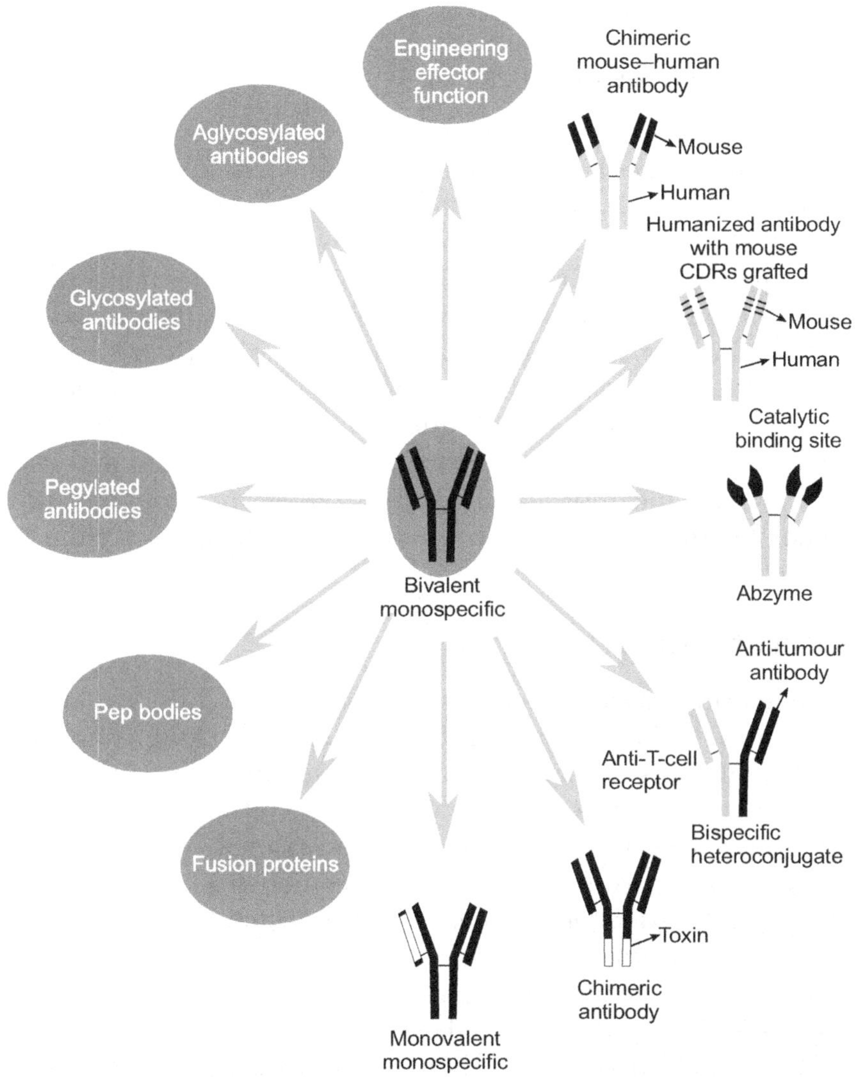

Figure 4.6 Diagrammatic representation of engineering of monoclonals. (modified from Kuby, J., 1997)

Chimeric and Humanized Antibodies

These antibodies are obtained by the homologous recombination in hybridoma cells as mentioned earlier, or are produced directly from a transgenic mouse. A better type of chimeric antibodies is produced from a monkey, known as **primatised antibodies**. This is more homologous to human than murine sequences and have proved to be less immunogenic.

Catalytic Monoclonals

The stereochemical and bimolecular interaction with no covalent chemical changes between the epitope and paratope of an antibody is similar in many ways to the binding of an enzyme to its substrate. In both the cases, the binding involves weak, non-covalent interaction with high specificity and affinity. These similarities made Lerner and his colleagues to discover an antibody with catalytic activity which they called as abzyme or catmab—a molecule with dual role (functioning both as antibody and enzyme). They prepared antibody to a transition state analogue, which exhibited thousandfold enhancement of catalysis with the kinetic patterns and substrate specificity typical for an enzyme. Abzymes are likely to represent a major technical advance that will have an impact on the various branches of science in future.

Bispecific Monoclonals

Two hybridomas are fused to obtain a hybrid hybridoma also known as quadroma that codominantly expresses the H and L chains of both its parent. This cell allows random assembly of two different light chains with two different heavy chains, resulting in an antibody with two different specificities. This antibody is called as bispecific antibody (Figure 4.6). Recently an interesting recombinant DNA technology has been developed where Ig genes are transfected to a transfectoma cell that is already secreting an antibody. The resulting clones are selected and found to secrete bispecific antibody. These monoclonals can be used effectively in immunohistochemistry, radioisotope-based or enzyme-based serodiagnosis, radioimmunoimaging, toxin therapy, etc.

The bispecific whole IgGs are also engineered by designing specific protrusions or **knobs** on one CH_3 domain and a corresponding hole in the CH_3 domain of the other antibody. This is achieved by swapping a small amino acid at the interface with a large one to make the knob and vice versa to produce a complementary hole. A more attractive and alternative approach of bispecific antibodies is linking of two such Fab′ fragments or linking of two Fab fragments via the hinge cysteine residues. The production of bispecific scFv and bispecific diabodies are discussed below.

The bispecific monoclonal antibody with one arm binding to an infected viral cell or a tumour cell and another to a cell-killing agent like drug, toxin or effector cell, can be used for safe, rapid binding, neutralization and removal from the circulation of pathogenic and associated infectious diseases and tumour cells. Designed to direct and enhance the body's immune response to specific pathogens, bispecific antibodies have shown promising results in phase I and phase II clinical trials, leading in some cases to complete or partial responses in infected patients.

Conjugated Antibodies

Antibodies are conjugated with variety of substances like radioisotopes, enzymes, reporter groups or drugs for diagnostic and therapeutic purposes. The degree of substitution, stability of linkages and biological activity of conjugates are the major issues of concern in the preparation of conjugates. Care must be taken to avoid loss of efficiency of CDR region or effector functions.

The phenolic hydroxyl group of tyrosine, carboxyl groups of aspartate and glutamate, amino group of lysine and thiol group of cysteine in the antibody are used for chemical modification and coupling.

Conjugation is carried out at specific sites that are away from the antigen-binding site, in order to avoid the loss of avidity. It can be performed by the

- modification of hinge cysteine residues.
- introduction of surface cysteine in constant region domains by protein engineering.
- modification of Fc carbohydrates.

- introduction of glycosylation site to a light chain by protein engineering.
- introduction of extra lysine residue by protein engineering for carbohydrate attachment.
- addition of C-terminus of antibody fragment by reverse proteolysis.

Fusion (Chimeric) Proteins

Another attractive and alternative category of monoclonal modification is the recombinant or fusion protein monoclonals. They are produced by fusing antibody or antibody fragment to any protein at the gene level. Expression of the first functional fusion protein was shown by the fusion of Fab fragment to staphylococcal nuclease. Since then the production of active antibody and antibody fragment with a variety of different proteins has been acheived. These fusion proteins have novel effector functions and have enormous clinical benefits. The proteins that are used in fusion proteins include enzymes like staphylococcal nuclease, β-lactase, β-glucuronidase, alkaline phosphatase and urokinase; toxins like pseudomonal exotoxin, gelonin, angiogenin and eosinophil-derived neurotoxin; cytokines and growth factors like IL2, GM-CSF, TNF-α and TNF-β, IGF-1 and IGF-2; and other proteins like aequorin, avidin, biotin, transferrin, protein A fragment and metallothionein, etc.

Monovalent Monospecific

They are formed when one homologous and one heterologous heavy- and light-chain association occurs resulting in one normal Fab arm and one inactive Fab arm respectively.

Pegylated Antibodies

To reduce the immunogenicity of monoclonals, polymers such as polyethylene glycol (PEG) or low-molecular-weight dextrin are added. Pegylated antibodies are obtained by attaching PEG via cyanuric chloride or tresyl chloride, or via N-hydroxysuccinimide. PEG conjugation to antibodies also changes the other biological functions such as resistance to proteolysis and increased circulating half-life. A human IgG–PEG conjugate with 20 PEG molecules per antibody, administered to mice was able to prevent 85–90% of immune response to subsequently administered unmodified human IgG.

Pepbodies

Pepbodies are obtained by a novel technology based on the fusion of small peptides bound to natural immunoglobulin effector ligands with antibody fragments such as Fab or scFv. Depending on the antibody specificity, pepbodies may be utilized as therapeutic agents by taking advantage of the body's own elimination methods such as the complement system or cellular Fc receptor-mediated systems. Pepbodies can recruit natural antibody effector functions without the need of an antibody Fc-region or additional antigen-binding domains. It can be constructed to combine more than one effector function by displaying different peptides as fusion proteins to the antibody fragment. It can discriminate between various effector functions and can differ from what is naturally associated with immunoglobulins. They have small size compared to intact antibodies and thus may facilitate tissue penetration. Active pepbodies can be produced in *E. coli*.

Aglycosylated Antibodies

Aglycosylated antibody is obtained either by the removal of N-linked carbohydrate attachment site of human IgG by site-directed mutagenesis or by the substitution of the Asn residue at position 297. Such antibody failed to exhibit any ADCC (antibody-dependent cell-mediated cytotoxicity) activity, but a significant level of CDC (complement-dependent cytotoxicity) activity is retained. When compared with the native antibody, aglycosylated antibody shows that not all effector functions of a human IgG are affected. The reduction in half-life of aglycosylated antibody appears to be dependent on isotype. Aglycosyl, chimeric IgG4 and IgG3, and murine IgG2b are cleared faster from blood as their half-life is reduced, whereas such chimeric IgG1 is unaffected as its half-life remains the same.

Glycosylated Antibodies

The researchers at GlycoFi and Dartmouth College have first reported the production of monoclonal antibodies with human sugar structures in yeast. This research work published online on January 22, 2006 and in the February 2006 issue of the journal *Nature Biotechnology*, demonstrates that antibodies with human sugar structures (glycosylation) can be produced in glyco-engineered yeast cell lines and that by controlling the sugar structures of antibodies, their therapeutic potency can be significantly improved. Moreover, this same approach offers the potential to improve other glycosylation-dependent drug properties such as solubility, half-life, or tissue distribution. Table 4.1 shows the protein-engineered therapeutic monoclonals that reduce the immunogenicity in patients.

Table 4.1 Types of monoclonals obtained by protein engineering

Type of antibody	Nature
Chimeric antibodies	Antibodies with mouse constant region and human variable region.
Humanized antibodies	Human antibodies with mouse CDR regions grafted.
Catalytic monoclonals	Antibodies with catalytic (enzymatic) sites.
Bispecific monoclonals	Antibodies with two different antigenic specificities.
Conjugated antibodies	Antibodies conjugated with toxin, drug, isotopes, etc.
Fusion (chimeric) proteins	Recombinant monoclonals with fusion protein.
Pegylated antibodies	Antibodies with PEG.
Monovalent monospecific	Antibodies with a normal Fab arm, and an inactive Fab arm.
Pepbodies	Antibody fragments with fused peptide.
Aglycosylated antibodies	Antibodies with no N-linked carbohydrate attachment site.
Glycosylated antibodies	Antibodies with engineered sugar structures.
Engineered antibodies for effector function	Antibodies with tailored effector functions.

Antibodies with Engineered Effector Functions

For binding Fc receptors and eliciting ADCC, human IgG1 and IgG3 are the most effective, whereas IgG2 or IgG4 can be chosen to minimize the Fc binding and ADCC effect. Human IgG1 and IgG3 molecules are most active for the complement activation. Most mouse and rat antibodies are poorly effective for the activation of human effector functions, although mouse IgG2a and rat IgG2b are more effective.

For therapeutic applications, the natural effector functions can be tailored according to the need. The residues from Leu 234 to Ser 329 in the lower hinge/CH_2 domain are responsible for Fc receptor binding. Mutations in these residues have been shown to reduce Fc receptor binding. In a humanized IgG4, when Leu-235 is changed to Glu, Fc receptor binding is lost. It must be remembered that mutations in these residues either may have unexpected effects on the other functional properties, or may increase the immunogenicity of the molecule, and therefore careful testing of the engineered molecule is required.

IgG1 exhibits the highest avidity because of its high flexibility and IgG2 exhibits the lowest avidity with its low flexibility. Thus the flexibility of the hinge region may affect the binding properties to cell-surface antigens even though the affinities are same. The core hinge region of human IgG4 contains the CPSC sequence, compared to the CPPC sequence of IgG1. The serine region present in the IgG4 sequence is responsible for increased flexibility. The altered isotype IgG4P that has been used for several chimeric or engineered human antibodies is obtained by changing the serine residue to proline in the IgG4 sequence.

It is possible to produce recombinant antibodies of other classes which are functional in Fc receptor binding. The mouse–human chimeric IgM, IgA1, IgA2 and IgD molecules are expressed and these molecules show functional binding to their respective receptors. The properties of these constant regions are therefore also available for incorporation into recombinant antibodies if required. Chimeric IgA2 is used to translocate across an epithelial cell layer to target tumour cells whose antigens are abundant on the apical cell surface.

The manipulation of antibody carbohydrate structure may therefore represent another route towards engineering antibody effector functions. Removal of carbohydrate by the mutation of the normal attachment site for N-linked carbohydrate at position 297 in the CH_2 domain results in severe reductions in the ability of antibody to mediate effector functions. The first approach to alter effector function is the production of polymeric IgG antibodies. Substitution of a serine residue near the C-terminus of the CH_3 domain (Ser 444) to cysteine allowed the production of IgG-dimer of a chimeric human IgG1, which is 200-fold more potent in complement-mediated lysis and ADCC. It can also be internalized like an IgA dimer and also offers advantages for delivery of fused cytotoxic agents into the target cell.

The second approach investigated is the expression of IgG molecules with tailpiece from the IgM chain which results in the production of pentameric or hexameric units with increased ability of complement activation and interaction with FcγRII receptor. Another approach is the use of bispecific antibody fragments to redirect effector functions. Bispecific diabodies with specificity for target antigen at one site and for serum IgG at the other site can retarget the serum IgG to target antigen and elicit complement activation, phagocytosis and ADCC.

MONOCLONALS FROM TRANSGENIC PLANTS

One of the most promising methods of producing antibodies is the use of transgenic plants. The plant cells in differentiated or dedifferentiated states are grown in controlled conditions, harvested from the biomass, purified easily, and transferred to the consumer in a minimal amount of time. Plants are the safe and cost-effective delivery system for many high-value products for today's market-reaching potential. Transgenic plants represent an economical alternative to fermentation-based production systems (Table 4.2). Plant-made antibodies (plantibodies) are especially striking, as plants are free from human diseases, thus reducing the screening costs of viruses and bacterial toxins.

Table 4.2 Comparison of production level and cost of monoclonals by different sources (Fernandez *et al.*, 2002)

Source	Production level	Cost in US $
E.coli	200 mg/L	
Mammalian cell lines	0.5 g/L	350–1200
Transgenic plants	500 kg/year/ 100–500 acres	80–250
Transgenic animals	8 g/L of milk	250

MONOCLONALS FROM TRANSGENIC ANIMALS

The production of monoclonals in the milk of transgenic animals is also a potentially low-cost route to the production of antibodies, which is currently under development. Transgenic sheep, goats and cows are being investigated for their ability to produce large volumes of milk containing the expressed protein. A high-level expression of functional antibody in transgenic goats has been reported, with levels of 10 g/litre produced in some animals. Recently it has been reported that monoclonal antibodies are expressed efficiently in the hen's egg.

MONOCLONAL-BASED GENE/CELL THERAPY/*IN VIVO* PRODUCTION

The production of a desired therapeutic antibody *in vivo* represents an elegant means of overcoming technical complications such as the short half-life of an injected product. A number of research groups are pursuing this strategy of gene therapy. This would also allow the expression of antibodies within individual cells, allowing the antibodies to target and hence inactivate (or perhaps activate) specific intracellular molecules and processes relevant to disease progression. By the attachment of appropriate cellular signal sequences, such intracellular antibodies (intrabodies) can target not only the cell cytoplasm but also specific intracellular organelles.

To render a long-term, cost-effective, infusion-free, mild, continuous, controlled and ectopic immunotherapeutic response, several methods have been achieved. They are

i. grafting of *ex vivo* modified myoblasts, skin fibroblasts and skin patches.
ii. intravenous and intramuscular injection of recombinant adenoviral vectors
iii. intramuscular injection of AAV (adeno-associated virus) vectors and
iv. implantation of monoclonal-producing cells encapsulated in an immunoprotective matrix made up of cellulose sulphate.

Gene Transfer Through Constitutive Expression Vector

The cDNAs for heavy chains (hTg10) and light (κTg10) chains of the Tg10 mouse monoclonal (IgG2a/κ) are cloned downstream of a CMV promoter in the pcoR-hTg10 and pcoR-κTg10 plasmids respectively. Mice are subjected to the intramuscular electroporation with an equimolar mixture of pcoR-κTg10 and pcoR-hTg10 and 5 others taken as negative controls with a saline solution. Tg10 levels in the bloodstream are subsequently followed up as a function of time. No Tg10 will be detected in control mice whereas all other mice expressed it for at least 142 days. Although variations between animals are observed, it is noteworthy that three of them express more than 1 μg/ml in the initial production found under non-optimized experimental conditions.

Regulatable Expression

It is obtained with multiple copies of bacterial tetracycline operator (tet O). Expression is controlled by tetracycline-controlled transactivator (tTA).

Direct DNA Transfer

The DNA carrying the engineered antibody gene can be introduced for therapy, either by encapsulating in a liposome, or by binding tightly to polycationic substances such as polylysine.

All these approaches have advantages and disadvantages of their own, and at present it is too early to say which is most successful in achieving target delivery *in vivo*. All of these approaches may be successful at low-efficiency gene transfer but high-efficiency delivery has not been achieved.

The ease of eternal generation of monoclonals with phage display and combinatorial library and advances made in protein engineering, and transgenesis in animals and plants, help us to get an array of more cheap, potential, multifunctional, less-immunogenic therapeutic molecules. Many pharmaceutical companies are hopefully involved in the clinical trials of these molecules as the future drugs.

REFERENCES

Burton, D. and Barbas, C.F. (1994). "Human antibodies from combinatorial libraries." *Advances in Immunology*. **57**: 191–280.

Carlsson, R., Martensson, C., Kalliomäki, S., Ohlin, M. and Borrebaeck, C.A.K. (1992). "The SCID mouse-A model to generate a human immune response and to produce human monoclonal antibodies." *J. Immunol.* **148:** 1065–1071.

Elbehri, A. (2005). "Biopharming and the food system: Examining the potential benefits and risks." *AgBioForum*. **8(1)**: 18–25. Available on the World Wide Web: http://www.agbioforum.org.

Glick, B.R. and Pasternak, J.J. (1998). *Molecular Biotechnology–Principles and Applications of Recombinant DNA.* ASM Press, Washington DC.

Green, L.L., Hardy, M.C., Maynard-Currie, C.E., Tsuda, H., Louie, D.M., Mendez, M.J. *et al.* (1994). "Antigen specific human monoclonal antibodies from mice engineered with human Ig heavy and light chain YACS." *Natl. Genet.* **7**: 13–21.

Houghton, A.N., Brooks, H., Cote, R.J., Taormina, M.C., Oettgen, H.F. and Old, L.J. (1983). "Detection of cell surface and intracellular antigens by human monoclonal antibodies. Hybrid cell lines derived from lymphocytes of patients with malignant melanoma." *J. Exp. Med.* **158**: 53–65.

Hudson, P.J. and Souriau, C. (2003). "Engineered antibodies." *Nature Medicine*. **9**: 129–134.

James, K. and Bell, G.T. (1987). "Human monoclonal antibody production, current status and future prospects." *J. Immunol. Methods*. **100**: 5–40.

Jones, P.T., Dear, P.H., Foote, J., Neuberger, M.S. and Winter, G. (1986). "Replacing the complementarity-determining regions in a human antibody with those from a mouse." *Nature*. **321**: 522–525.

Kara, A.E., Bell, C.W. and Chin, T.E. (1995). "Recombinant antibody technology." *ILAR Journal*. **37**: 132–141.

Karpas, A., Dremuchera, A. and Czepulkowski, B.H. (2001). "A human myeloma cell line suitable for the generation of human monoclonal antibody." *Proc. of Natl. Acad. Sci.* USA. Vol. **98**: No.4, 1799–1804.

Larrick, J.W. and Bourla, J.M. (1986). " Prospects for the therapeutic use of human monoclonal antibodies." *J. Biol. Response Mod.* **5**: 379–393.

Lewis, A.D., Chen, R., Montefiori, D.C., Johnson, P.R. and Clark, K.R. (2002). "Generation of neutralizing activity against human immunodeficiency virus type 1 in serum by antibody gene transfer." *J. Virol.* **76**: 8769–8775.

Mc Cafferty, J., Griffiths, A. D., Witer, G. and Chiswell, J. D. *et al.* (1990). "Phage antibodies: filamentous phage displaying antibody variable domains." *Nature.* **348**: 552–554.

Morrison, S., Johnson, M.J., Herzenberg, A.L. and Oi, T.V. (1984). "Chimeric human antibody molecules: mouse antigen-binding domains with human constant region domains." *Proc. Natl. Acad. Sci.* USA. **81**: 6851–6855.

Noel, D., Pelegrin, M., Brockly, F., Lund, A.H. and Piechaczyk, M. (2000). "Sustained systemic delivery of monoclonal antibodies by genetically modified skin fibroblasts." *J. Invest. Dermatol.* **115**: 740–745.

Noel, D., Pelegrin, M., Kramer, S., Jacquet, C., Skander, N. and Piechaczyk, M. (2002). "High *in vivo* production of a model monoclonal antibody on adenoviral gene transfer." *Hum. Gene. Ther.* **13**: 1483–1493.

Noel, D., Dazard, J.E., Pelegrin, M., Jacquet, C., and Piechaczyk, M. (2002). "Skin as a potential organ for ectopic monoclonal antibody production." *J. Invest. Dermatol.* **118**: 288–294.

Ohlin, M., Kristensson, K., Carlsson, R. and Borrebaeck, C.A.K. (1992). "Epstein–Barr virus-induced transformation of human B lymphocytes-The effect of L-leucyl-L-leucine methyl ester on inhibitory T cell populations." *Immunol. Lett.* **34**: 221–228.

Owens, R.J. and Young, R.J. (1994). "The genetic engineering of monoclonal antibodies." *Immunol. Methods.* **168**(2): 149–165.

Pasqualini, R. and Arap, W. (2004). "Hybridoma-free generation of monoclonal antibodies." *Proc. Natl. Acad. Sci.* USA. **101**: 257–259.

Pelegrin, M., Marin, M., Noël, D., Del Rio, M., Saller, R., Günzburg, W., Stange, S., Steffen, M. and Piechaczyk, M. (1998). "Systemic long-term delivery of antibodies in immunocompetent animals using cellulose sulphate capsules containing antibody-producing cells." *Gene Therapy.* **5**: 828–834.

Perez, N., Bigey, P., Scherman, D., Danos, O., Piechaczyk, M. and Pelegrin, M. (2004). "Regulatable systemic production of monoclonal antibodies by *in vivo* muscle electroporation." *Genetic Vaccines and Therapy.* **2**: 2.

Rechmann, L., Cark, M., Waldman, H. and Winter, G. (1988). "Reshaping human antibodies for therapy." *Nature.* **332**: 323–327.

Suresh, M.R., Noujaim, A. and Longenecker, B.M. (1991). "Recent development in monoclonal antibodies." In: *Biotechnology–Current Progress.* Vol. I. Cheremisinoff, P.N. and Ferranta, L.M. (eds.). Academic Publishing Co. Nc. Lancaster.

Winter, G., Griffiths, A., Hawkins, R. and Hoogenboom, H. (1994). "Making antibodies with phage display technology." *Ann. Rev. Immunol.* **12**: 433–455.

Wright, A., Shin, S.U. and Morrison, S.L. (1992). "Genetically engineered antibodies-progress and prospects." *Crit. Rev. Immunol.* **12** (3–4): 125–168.

Wright, A. and Morrison, S. (1997). "Effect of glycosylation on antibody function: Implications for genetic engineering." *Trends Biotechnol.* **15**: 26–32.

5 MONOCLONALS FROM TRANSGENIC PLANTS AND ANIMALS

The transgenic plants, first reported in 1983, offer a new approach to the production of human monoclonals. Monoclonals have been expressed in several important agronomic species of plants including tobacco, corn, tomato, potato, banana, rice, wheat, maize, oilseed rape, alfalfa and canola. The production of protein biologicals in recombinant plant systems is called **phytopharming**. This phytopharming has shown great promise in the studies performed over the past few years.

MONOCLONALS (PLANTIBODIES) FROM TRANSGENIC PLANTS

To produce recombinant monoclonals in plants, several transformation approaches are used. They are produced by the following methods.

1. Delivering the antibody genes directly through gene gun technology.
2. Transferring the genes through *Agrobacterium tumefaciens*, where the antibody-producing genes are integrated into the plant genome, thereby guaranteeing perpetual propagation.
3. Infecting the transgenic plants with modified recombinant viruses, thereby expressing the antibody genes during their replication in the host.

METHODS ADOPTED

After transformation, the pharmaceutical and therapeutic plantibodies can be produced in a variety of ways.

In vitro Cell Tissue Cultures

Transformed cells, as they are totipotent, can be propagated indefinitely in tissue cultures. The transgene stability is increased in this type of asexual propagation due to the absence of crossing over, segregation and recombination. The problems associated with open-field production can be solved by inducible promoters with better efficiency and stable mode of translation, leading to stability of the transgene.

Breeding and Sexual Crossing

In 1989, Hiatt *et al.* produced functional antibodies in plants, by introducing the kappa-chains of either light or heavy regions into one tobacco plant and the gamma-chains of either light or heavy regions into another plant and upon crossing them, they produced antibody that expressed both the chains in an ingenious way, without the need for a double transformation.

Expressing the Genes in Seeds (or) Seed Production System

Plants cannot store antibodies for an extended period of time due to protease degradation whereas seeds contain a low level of proteases, which allows proteins to be stored without degradation. Proteins stored in the seeds remain desiccated and have been shown to retain their potency and quantity intact. Hence seeds can be used as bioreactors and as natural storage organs. For this the antibody genes are directed to be expressed in a tissue-specific manner in the seeds of crops such as corn, rice, wheat, soybeans and rape seed. An inbred corn plant yields 100–5000 seeds in 4–6 months time, each seed weighing about 0.2 g. Corn is thus a better seed production system comparable to other plants.

Targeting and Compartmentalization

In order to obtain high production of antibody, it has been found that targeting the proteins into the extracellular space (i.e., the apoplasm) is the most efficient method. Plants, like animals, secrete antibodies after post-translational modification has occurred. Proteins that are secreted into the apoplastic space undergo less hydrolysis due to the lack of hydrolytic enzymes in the apoplasm. Signal peptides can be used to target the light and heavy chains to the endoplasmic reticulum (ER). Chaperones similar to those found in human cells are found here, which ensure the correct folding of an antibody. A higher accumulation of antibody fragments is observed in the ER and the apoplast compared to the cytosol.

Targeting and compartmentalizing to easily isolatable organelles provides a less-complicated purification procedure. Functional antibodies can also be harvested from transgenic potato tubers and even from seeds, where they can be safely stored for long periods of time. The highest yields in seeds have been achieved after retention in the endoplasmic reticulum. Targeting, however, has to be specifically controlled, so that cleavage of the targeting sequence during purification is easy. If incomplete processing occurs, the quality and the amount of protein will be lowered.

TYPES OF PLANTIBODIES

The following are the successfully achieved monoclonal plantibodies.

Secretory IgA

Julian *et al.*, (1994) produced a functional secretory (sIgA) antibody for a cell surface protein of *Streptococcus mutans* that causes dental caries in man, by sexually crossing four plants, expressing each of the polypeptides of antibody dimer. sIgA not only contains heavy and light chains but it is also dimerized by a J chain, and protected from proteolysis by a fourth polypeptide, the SC (secretory component).

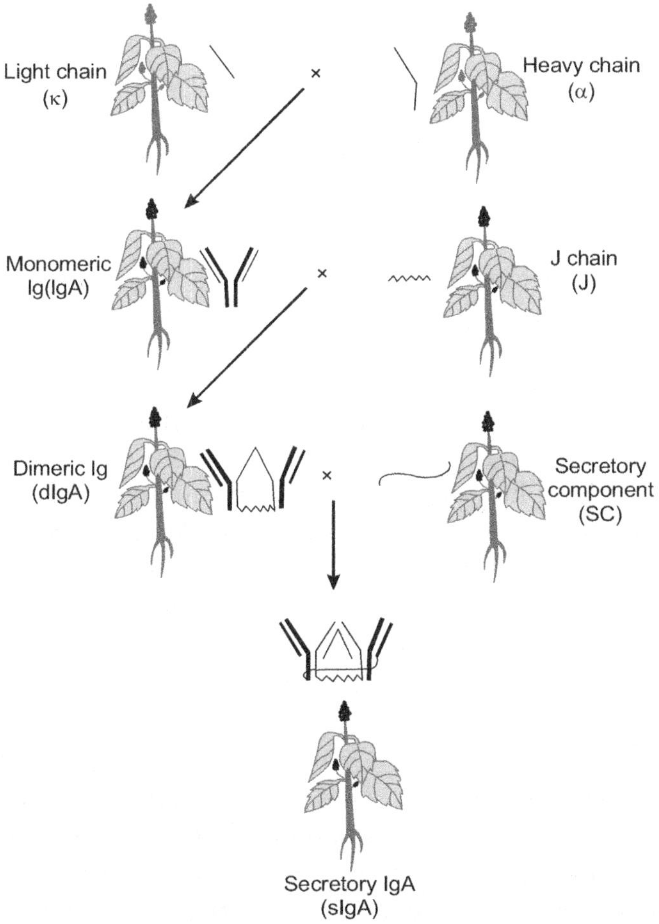

Figure 5.1 Strategy for production of secretory antibody in plants (Sharma, A.K., 1999)

A transgenic plant carrying the light-chain gene is crossed with another transgenic plant carrying the heavy-chain gene. This hybrid transgenic plant can produce a complete IgA monomer which in turn is crossed with another transgenic plant carrying the J chain. The second hybrid plant is capable of producing IgA dimer (Figure 5.1). The transgenic plant that is capable of secreting IgA dimer is crossed with the fourth transgenic plant capable of producing secretory component. The resulting hybrid is capable of producing secretory IgA. It is screened by ELISA using SA I/II antigen, a cell surface protein of *Streptococcus mutans,* that causes dental caries in humans.

Larrick *et al.* (1998) orally administered to 84 human subjects, CaroRx™, an anti-*S.mutans* sIgA, a clinically advanced sIgA plantibody from transgenic tobacco. By topical (passive immunotherapy) applications, it prevented the colonization of artificially and naturally implanted *S. mutans,* and this protection from recolonization (with just 3 weeks of application) lasted for two years.

Hybrid IgA/G

A hybrid monoclonal antibody (IgA/G), having constant regions of IgG and IgA fused, has been used successfully against human dental caries caused by the bacterium *S. mutans*. Ma *et al.* compared the secretory antibody generated in transgenic tobacco (sIgA/G) and the original mouse IgG. Though both had similar binding affinity to the surface adhesion (SA) protein of *S. mutans,* sIgA/G could survive for 3 days in the oral cavity, whereas IgG could survive for just one day. The purified antibody applied over a period of three weeks to the teeth of volunteers, after treatment with chlorhexidine, succeeded in preventing the colonization of *S. mutans* in oral mucosa for up to four months. If the current large-scale trials are successful, it is intended to include this antibody as a component of toothpaste.

Fab or scFv

Ulrike Fieldler and Udo Conrad (1995) used the seeds of transgenic tobacco for high-level production and long-term storage of antibodies in the form of Fab or scFv and the leaves of tobacco for high-level expression. (Tobacco yield of seeds per plant is high (11,000 seeds/g).) In their experiments, functionally active scFv accumulated up to 0.67% of the total soluble seed proteins. Storage of ripe transgenic tobacco seeds for one year at room temperature showed no loss of scFv and its antigen-binding activity. The level of expression in the tobacco leaves has been reported to be between 0.01% and 6.8% of the total soluble protein.

Stoger *et al.* produced an scFv recombinant fragment against carcino-embryonic antigen (CEA) a well-characterized tumour-associated surface antigen, in transgenic rice plants and cell lines. Three different scFv constructs were engineered to optimize the product yields, allowing monoclonal targeting to different cell compartments. An N-terminal murine leader sequence was encoded by all the three constructs (Figure 5.2) to target the scFvs to the secretory system.

1. The construct A encoded a C-terminal KDEL retention signal, to retain scFvs in the endoplasmic reticulum (ER).
2. The construct B encoded a C-terminal His_6 tag instead of the KDEL signal, and antibodies passed through the secretory system to the apoplast.
3. Transgenic rice plants were generated with each of these constructs driven by two alternative promoters, the enhanced CaMV 35S promoter (construct C) and the maize ubiquitin-1 promoter (Figure 5.2).

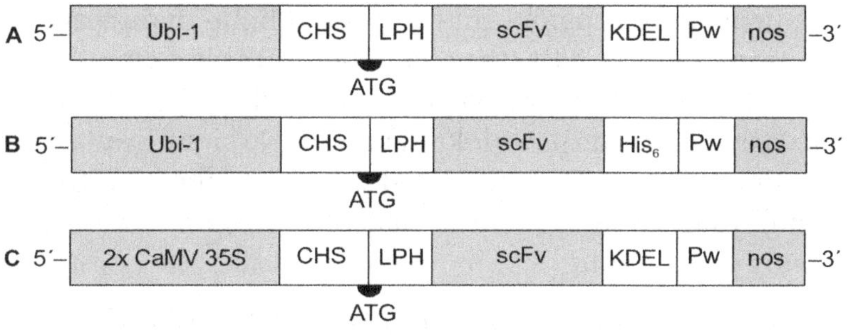

Ubi-1—Maize ubiquitin promoter + intron
2x CaMV 35S—Enhanced cauliflower mosaic virus 35S promoter
CHS—Chalcone synthase 5′-untranslated region
LPH—Heavy-chain leader peptide from TMV virion-specific mAB24*
scFv—T84.66 scFv-coding region
KDEL—Endoplasmic reticulum retention signal
His_6—Hexameric histidine tag
Pw—TMV 3′-untranslated region
nos—nos termination sequence

Figure 5.2 Tailoring of three different constructs of scFv against CEA in rice (Stoger, *et al.*, 2002)

Competitive ELISA assays were used to determine the yield of functional recombinant antibody. Up to 30 μg/g fresh weight of antibody was detected in the leaves and seeds of transgenic rice plants. The highest expression levels were achieved using the KDEL construct driven by the enhanced CaMV 35S promoter.

They also explored the viability of rice cell lines (Figure 5.3) derived from transformed callus, as a system for stable and sustainable antibody production. In these experiments, they investigated the potential of a TMV omega sequence and chalcone synthase gene constructs in the 5´-untranslated region (UTR) and leader sequence. They compared alternative constructs encoding C-terminal His_6 tag and KDEL retention sequences, with all constructs driven by the maize ubiquitin-1 promoter.

Levels of functional antibody production were determined by competitive ELISA assay. The highest production level, (approximately 3.5 μg/g callus fresh weight) was achieved using the KDEL expression construct. The antibody levels were 6–14 times higher in cells transformed with the KDEL construct compared to those transformed with the His_6 construct. Also they generated an extended series of KDEL constructs with all possible combinations of alternative leader peptides (the murine immunoglobulin heavy- and light-chain leader peptides) and alternative 5´-UTRs (the chalcone synthase gene 5´-UTR and the tobacco mosaic virus omega sequence). The 3´-UTR comprised the tobacco mosaic virus pseudoknot region. No significant differences in the antibody levels were recorded in each of the resulting cell lines, showing that these sequences had only little effect on the level of antibody production.

From this observation, they have drawn the following conclusions:

- The rice cell line culture system offers a constant and sustainable antibody production.
- The transgenic plants are easy to propagate and harvest, and the seed can be used as a long-term storage vehicle.
- The rice transgenic system could emerge as an important and beneficial production system for recombinant pharmaceuticals and other valuable products.

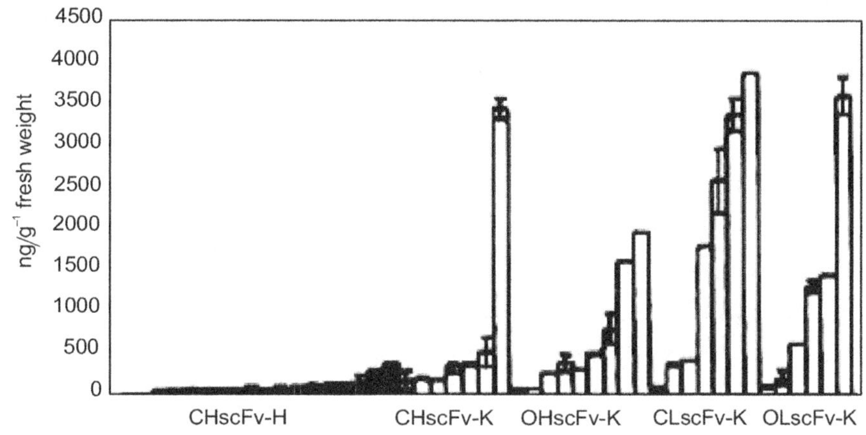

CHscFv-H—Solid bars are cell lines containing His_6 tag and not KDEL signal.

CHscFv-K—Chalcone synthase gene 5′ -UTR and murine IgG heavy-chain leader peptide.

OHscFv-K—TMV omega sequences 5′-UTR and murine IgG heavy-chain leader peptide.

CLscFv-K—Chalcone synthase gene 5′-UTR and murine IgG light-chain leader peptide.

OLscFv-K—TMV omega sequence 5′-UTR and murine IgG light-chain leader peptide.

Figure 5.3 Levels of scFv production in rice cell lines. Bars are representing individual cultures expressing one of the constructs (Stoger, *et al.*, 2002)

scFvs, unlike full-length antibodies, can easily be targeted to sub-cellular compartments. The plant breeder has been exploiting this property in targeting specific pathogen proteins in plants, in order to engineer pathogen resistance. Few such scFvs are presented in Table 5.1.

Table 5.1 Antibodies produced *in situ* in transgenic plants

Antigen	Plant	Antibody	Application
Artichoke mottled crinkle virus coat	*Nicotiana*	scFv	Viral protection
Beet necrotic yellow vein virus	*Nicotiana*	scFv	Viral protection
Root knot nematode	*Nicotiana*	scFv	Nematode protection
Fungal cutinase	*Nicotiana*	scFv	Fungal protection
Abcisic acid	*Nicotiana*	scFv	Wilty phenotype

IgG

Full length antibodies are advantageous than the fragments of antibodies because certain domains could be altered to facilitate complement activation upon binding. This would significantly aid the therapeutic properties of the antibody if it has to be used in the treatment of human disease, helping to elicit an immune response.

It has been possible to raise IgG antibodies to virtually any antigen in the laboratory. IgG antibodies have been expressed at levels of 0.35 to 1.3% of the total soluble protein in tobacco leaves. The IgG1 plantibodies fulfil many of the qualities that are desirable for vaginal microbicides—efficacious, specific, and potent molecules. In a study evaluating an anti-HSV monoclonal, it was found to be 100–1000 times more potent on a weight basis than the other agents tested for preventing vaginal transmission of genital herpes infection in mice. These results show that this strategy could be useful for many other mucosal infections in humans and animals.

Because of their potency, these monoclonals can be applied in small volumes, allowing women-controlled protection that is undetectable to their partners. Because of their vaginal residence time, monoclonals may provide protection even on the days when the woman fails to use the method. For longer-term protection, monoclonals can also be delivered by sustained release from intravaginal devices.

Therapeutic Plantibodies

With the growing fulfillment, the clinical potentials of plantibodies are likely to allow development of inexpensive monoclonal-based health-care products. The chimeric antibody against hcg has been transiently expressed in tobacco leaves. The yield of the purified antibody was 20–24 mg per kg of fresh leaves of tobacco. Probing experiments have shown that several non-tobacco plants such as spinach, zuchini, aubergines, etc. can also make this antibody. Rapid production of patient-tailored anti-lymphoma antibodies in recombinant tobacco (virus-infected tobacco) may provide an effective cancer therapy. Table 5.2 lists out a few plantibodies generated to treat infectious diseases and cancer.

Table 5.2 Pharmaceutical monoclonals produced in transgenic plants (*Trends in Plant Science.* May 2001, Vol.6 No.5, pp. 220.)

Monoclonal	Antigen	Transgenic plant	Application
Guy's 13 (sIgA)	Streptococcal antigen	*Nicotiana tabaccum*	To cure dental caries
C5-1 (IgG)	Anti-human IgG	Alfalfa	As diagnostic
scFvT84.66	CEA	Wheat, rice	Cancer treatment
T84.66 (IgG)	CEA	*N. tabaccum*	Cancer treatment
Diabody	CEA	*N. tabaccum*	Cancer treatment
38C13 (scFv)	Rice alpha amylase	*N. benthamiana*	Lymphoma treatment
CO17-1A	Surface antigen	*N. benthamiana*	Colon cancer
Anti-HSV-2 (IgG)	HSV-2-antigen	Soybean, rice	HSV-2 infection
RSV-IgG	RSV antigen	Corn	RSV infection (inhalant)
Sperm-IgG	Sperm	Corn	Contraceptive (topical)
IgG	Colon cancer antigen	*N. tabaccum*	Therapeutic/diagnostic
IgG	*Clostridium difficile*	Corn	Therapeutic (oral)

Thus, the transgenic plant technology presents a viable alternative for economical production of human therapeutic proteins. It is considered imminent that plant-derived human therapeutic proteins will be in the market in a few years. The "points to consider" and "guidance" published by the FDA for the transgenic-animal-produced therapeutics are being adapted for plant-derived therapeutics. If the purity of the plant-produced material can be maintained on par with that of mammalian cell-culture derived products, then the plant-derived products will have an advantage in terms of cost benefit.

BIOEQUIVALENCE

The corn-seed-derived antibody was found to be equivalent to the mammalian-derived antibody in biochemical properties, serum and urine clearance and effectiveness on target-tissue binding. The corn-seed-derived antibody met the purity, safety and potency requirements that are necessary for use in human clinical evaluation. It was also demonstrated that the seeds stored under controlled conditions provided viable seeds for at least seven years. In addition, the quality and yield of the antibody produced by milling the corn right after the harvest and after a storage period of 20 months remained the same. In fact, in a recently completed clinical trial with repeated applications of plantibodies for the prevention of oral colonization by *Streptococcus mutans*, no safety problems were encountered, nor were there any detectable human anti-plant antibody responses.

COST OF PLANTIBODIES

Monoclonals are produced in cell culture for $200 to $1,000 per gram. Production by transgenic plants indeed lowers the cost. Transgenic plants can be scaled up in agricultural fields to produce tonnes of plantibodies, and these plantibodies are predicted to cost less than US $1/g.

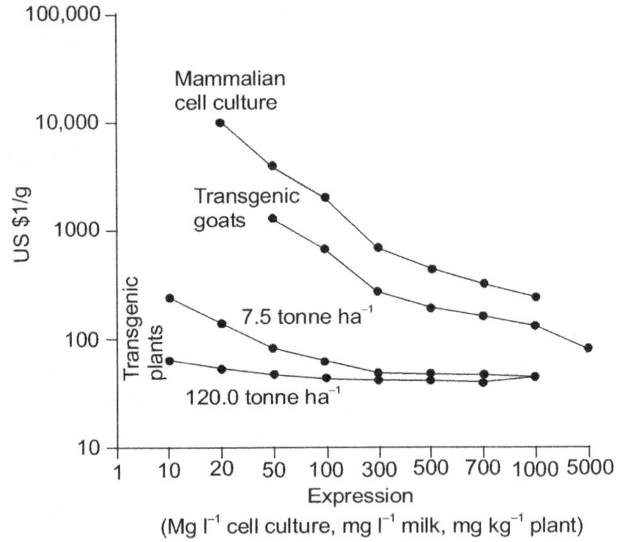

Figure 5.4 Graph showing the cost of purified IgA in different expression systems (Stoger, *et al.*, 2002)

The actual cost, however, will remain unknown until large-scale batches are produced, purified, and formulated in accordance with good manufacturing practices. A typical therapeutic protein costs $ 200/g of pure protein at a production level of 200 kg/year. If the same protein is produced in transgenic corn, the cost is expected to be about \$ 50/g at a production level of 100 kg of a monoclonal antibody. At a 100,000 kg/year level, the cost will be \$ 4.5/g. The major cost difference is in the upstream processing since the purification steps are the same in both cases.

Figure 5.4 gives the cost of purified IgA in different expression systems. Costs for mammalian cell culture are derived from industrial costs for cell culture and purification facilities. The cost for transgenic goats is derived from publicly available estimates from Genenzyme Transgenics. The cost for plants has been compared with green biomass (120 tonne ha^{-1} in the Figure 5.4) and seed production (7.5 tonne ha^{-1}). Differences in cost are based primarily on the production cost, because purification cost is the same for all systems.

ADVANTAGES AND DISADVANTAGES

Antibody production using transgenic plants as bioreactors is likely to have more economical and qualitative benefits than alternative systems, especially for large-scale needs. The following are the factors that are in favour of plant systems as sources of animal-derived proteins compared with other conventional methods.

- The potential for large-scale, low-cost biomass production using agriculture.
- The *ex vivo* stability and *in vivo* efficacy of these plantibodies have antigen-binding ability similar to the same proteins expressed in bacterial or mammalian expression systems.
- There is a reduced risk of contamination by prions, mammalian viruses, blood-borne pathogens, oncogenes and bacterial toxins.
- The yields can be high, and if expressed in seeds lead to convenient processing.
- The capacity of plant cells to correctly fold and assemble not only the antibody fragments and single chain peptides, but also the full-length multimeric proteins.
- Transgenic plants not only provide the means to express antibodies but also enable the glycosylation and entry into secretory pathway which allow the assembly of complete antibodies and Fab fragments. Variable fragments (Fv) can be produced in cytosol, directed to different compartments and

fused with proteins such as protein A and phosphatase to improve the detection and purification of single chain Fv (scFv).

- Low downstream processing requirements for the proteins administered orally. Elimination of the purification requirement when the plant containing the recombinant proteins is edible, such as potatoes.
- The ability to introduce new or multiple transgenes by sexual crossing of plants.
- The avoidance of ethical problems associated with transgenic animals.
- Formulated in seeds, plant-made enzymes have been found to be an extremely convenient method for reducing storage and shipping costs, for an indefinite amount of time, under ambient conditions.
- Production size is flexible and easily adjustable to the needs of changing markets.
- Plants are also capable of synthesizing and assembling virtually any kind of antibody molecule, ranging from the smallest antigen-binding domains and fragments, to full-length, and even multimeric antibodies.

There are, however, potential issues of concern for plantibodies production. They are the following.

- Allergic reactions to plant protein glycans and other plant antigens.
- Even though mammalian glycans are similar to plant glycans in size and extent of branching, the predominant terminal residue in mammals is N-acetyl neuraminic acid, a carbohydrate not present in plants.
- The glycosylation pattern is different and some carbohydrate moieties are unique to plants and may present an antigenic challenge to the immune system when administered on a regular basis.
- Contamination of the plant and its product by mycotoxins, pesticides, herbicides and endogenous metabolites.
- Regulatory rules are uncertain, particularly for plant proteins requiring approval for human drug use.

GENE SILENCING

Recent studies with transgenic production of antibodies have shown that the transgene involved can undergo inactivation; this process is termed gene silencing. Gene silencing has been observed to occur where there exist multiple-copy integrations at one or more sites, different base-compositions between rDNA and the integration site, detrimental effects of sequences adjacent to the rDNA integration site and over-expression effects. This is prevented by

- screening/selection for plants with single copy rDNA;
- developing methods for single-copy integration;
- avoiding repetitive homologous sequences;
- flanking rDNA with scaffold attachment regions;
- screening/selection for stable rDNA expression; and
- developing site-specific recombination systems.

MONOCLONALS FROM TRANSGENIC ANIMALS

The first transgenic animal, a lowly mouse, was created in Yale University (New Haven, CT, USA) in 1980. Since then many other transgenic animals, including sheep, goats, cows, rabbits, pigs and chicken, have been created at universities and corporate research laboratories. Transgenic mice provide a ready and alternative source of diverse, high-affinity, high-specificity and low immunogenicity therapeutic monoclonals.

PRODUCTION OF TRANSGENIC MICE

Alt *et al.* (1985) suggested that transgenic technology could be useful for generating new human sequence monoclonals starting from unrearranged, germ-line configuration of transgenes. The authors concluded that although this was "conceptually outlandish," it might "be realized in the not-too-distant future." In 1989, Bruggemann *et al.* reported the expression of a repertoire of human heavy chains and the generation of a transgene-encoded immune response in mice. This report and the invention of methods for introducing specific modifications into the mouse germ line fuelled the race to generate a mouse that comprised diverse human heavy- and light-chain repertoires capable of contributing to a true secondary immune response of high-affinity human monoclonals with low immunogenicity, in the background of disrupted mouse heavy- and κ-light-chain genes.

In 1994, two papers, one from Nils Lonberg of Medarex, (California, USA) and the other from Green *et al.* of Cell Genesys (Foster City, CA, USA), reported the generation of mice with four different germ-line modifications: Nils Lonberg used pronuclear microinjection to introduce reconstructed minilocus transgenes—the heavy chain containing 3 heavy-chain variable (V_H) regions, 16 diversity (D) regions and all 6 heavy-chain joining (J_H) regions together with μ and γ_1 constant-region gene segments. In the transgenic strains, this construct underwent VDJ joining, together with somatic mutation and correlated class switching. The light-chain transgene included four V_κ, all five J_K and the κ-constant region (C_κ).

In contrast, Green *et al.* used protoplast fusion to deliver the YAC-based (YAC—Yeast artificial chromosome) minilocus transgene. In this case, the heavy chain included 5 V_H, all 25 D and all 6 J_H gene segments together with μ and δ constant-region gene segments. This construct underwent VDJ joining and expressed both IgM and IgD. The light-chain YAC construct included two functional V_κ and all five J_κ segments, together with C_κ. Neither the construct of Nils Lonberg nor that of Green *et al.* inactivated the endogenous λ-light-chain locus, which in typical laboratory mouse strains contributes to only ~5% of the B-cell repertoire.

Since 1994, there have been multiple reports in the literature of transgenic mice. Several different technologies, including pronuclear microinjection and yeast protoplast fusion with ES cells (Figure 5.5), have been employed for engineering these mice strains. The introduction of the largest fraction of the human germ line repertoire is facilitated by one particular technological innovation: microcell-mediated chromosome transfer. In this approach, human fibroblast- derived microcells are fused with mouse ES cells resulting in pluripotent cell lines having a single human chromosome or a chromosome fragment—including a centromere and both telomeres—that replicates and assorts during cell division without insertion into an endogenous mouse chromosome.

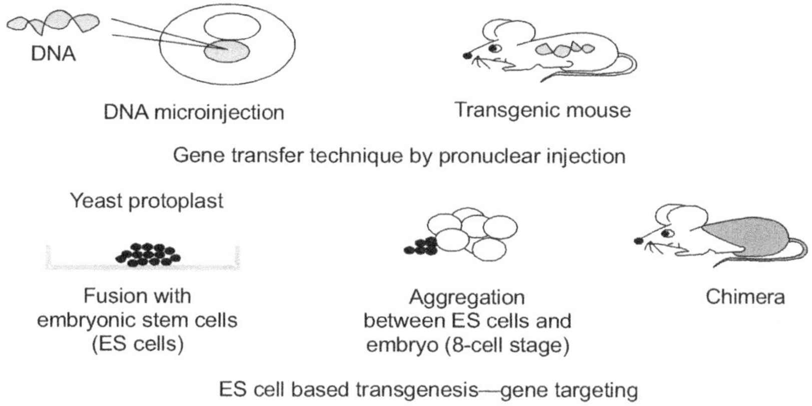

Figure 5.5 Production of transgenic mouse

Using this technique, Tomizuka *et al.* (2000) generated ES cell lines and chimeric mice containing fragments of human chromosomes 2 and 14, including the human κ-light-chain and heavy-chain loci, respectively. In addition, they generated chimeric mice containing an apparently intact human chromosome 22, comprising the λ-light-chain locus.

In a subsequent report, germ-line transmission was obtained with a human heavy-chain ES cell line, and mice were created that expressed complete human heavy- and light-chain repertoires in a genetic background that included disruptions of the mouse heavy- and light-chain loci. Completely human, high-affinity (< nanomolar) monoclonals were isolated from the animals. Although both the chromosome fragments could be transmitted through the germ line, the light-chain-containing chromosome-2 fragment was found to be less mitotically stable. The stability has been increased by deleting non-immunoglobulin genes between IgH and the centromere, resulting in 10 to 20 Mb fragments. This minimizes cross-species trisomy and increases selection during mitosis. This structure has now been exploited to create artificially constructed human chromosome fragments that include the entire human heavy-chain locus together with the entire human λ-light-chain locus.

A conceptually analogous transgenic system that generates chimeric antibodies, rather than fully human sequence antibodies, was developed by Rajewsky *et al.* (1994). Mice generated using this approach contain relatively precise replacements of the mouse κ and

γ_1 constant-region gene segments with the corresponding human gene sequences. The κ-constant-region gene segment was replaced using homologous recombination in mouse ES cells. For the γ_1 gene, only the secreted exons were replaced, and the engineering was accomplished in two steps using the *Cre-loxP*. The *Cre-loxP* recombination system is a genetic tool to control the site-specific recombination events in genomic DNA. *Cre* sequence is responsible for Cre protein, a DNA recombinase, that binds with specific *loxP* sequences. Using this technology, specific tissue type or cells of organisms can be genetically modified whilst other tissue remain unchanged.

In the 11 years since the first of these platforms was reported, a variety of different human sequence monoclonals derived from transgenic mice have been described in the scientific literature. These have included monoclonals against small molecules, pathogen-encoded protein and polysaccharide antigens, human-secreted and cell-surface protein antigens, and human tumour-associated glycosylation variants.

Clinical Development of Human Monoclonals from Transgenic Mice

At least 35 fully human monoclonals have entered clinical development as therapeutics. Two of these, α-interleukin (IL-8) (ABX-IL8 for psoriasis; Abgenix, Fremont, CA, USA) and αMuc18 (ABX-MA1 for melanoma; Abgenix, Fremont, CA, USA), have dropped out of development, whereas 33 remain in clinical trials. These molecules are targeted against a variety of clinical indications in cancer, autoimmune or inflammatory and infectious diseases. Four antibodies, directed against epidermal growth factor receptor (EGFR), receptor activator of nuclear factor-κB ligand (RANKL) also called osteoprotogerin ligand (OPGL), cytotoxic T-lymphocyte antigen 4 (CTLA-4), and CD4, are now in phase III development.

Immunogenicity of Human Monoclonals from Transgenic Mice

A review of the available clinical data gives us an opportunity to ask whether the transgenic mouse platforms have actually solved the problem of immunogenicity that originally motivated their development. Although transgenic-mouse-derived human monoclonals have yet to emerge from a phase III clinical study that could provide data comparable to those available for approved products, the initial results are encouraging.

Many existing approved therapeutic monoclonals from chimerization, CDR grafting and phage display, have been found to be immunogenic. In contrast, monoclonals from transgenic mice show no immune response. However there are exceptions. Adalimumab (Humira; Abbott Laboratories, Abbott Park, IL, USA), a phage display derived antibody directed against TNF-α and another phage display-antibody directed against IL-12 (the drugs used in inflammatory diseases) have been found to be immunogenic in human patients, despite the fact that they are derived from human rather than mouse sequences.

The transgenic technique also produces high-affinity human sequence monoclonals against a wide variety of potential drug targets. It is also directly used for the production of therapeutic human-sequence polyclonal antibodies.

PRODUCTION OF MONOCLONALS IN MILK

In the mid-1990s, the new area of nutraceuticals has been championed as a more attractive biotechnology area to invest in, than pharmaceuticals. Over other bioreactor technologies, the mammary gland bioreactor technology wins hands down for complex therapeutic proteins. This may be due to the cost of production, the authenticity of the protein product (post-translational modifications), or the cost of extraction and purification, or a combination of all the three. In many applications, a herd of several hundred cows or a flock of several thousand sheep might actually be capable of satisfying the world demand for a pharmaceutical and, for the most optimistic scenarios, the value of the pharmaceutical produced by an individual animal could be in the 10s to 100s of millions of dollars over the productive life of the animal. Thus, the mammary gland of a farm animal has been found to be an attractive way to go.

To make therapeutic monoclonals in the milk of transgenic animals, the procedures should make sure that the right control genes are put in besides the gene for the Ig and they should be directed to go to the mammary glands of the animal so that the protein will be expressed in the milk. The success rate is then measured by the extent of expression of the protein in the milk of the transgenic animal (Figure 5.6).

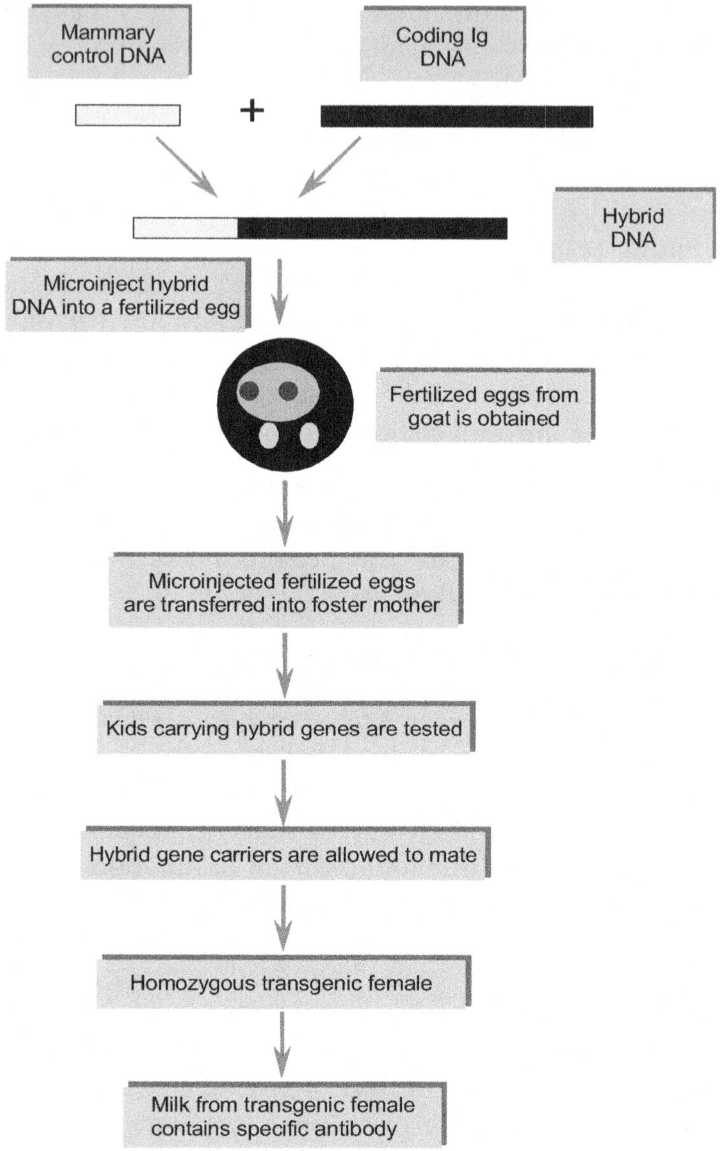
Mammary
control DNA
Coding Ig
DNA
+
Hybrid
DNA
Microinject hybrid
DNA into a fertilized egg
Fertilized eggs from
goat is obtained
Microinjected fertilized eggs
are transferred into foster mother
Kids carrying hybrid genes are tested
Hybrid gene carriers are allowed to mate
Homozygous transgenic female
Milk from transgenic female
contains specific antibody

Figure 5.6 Production of monoclonals in the milk of a transgenic goat

Transgenic mammals that secrete monoclonals in their milk have been generating one gram of antibody for roughly $100—one-third the cost of traditional production methods. Centocor, and Johnson and Johnson are looking into producing Remicade (a therapeutic monoclonal drug for breast cancer) using transgenic goats, and Infigen in DeForest, Wis., intends to make monoclonals in cow's milk, although no such products have yet received FDA approval.

Departing from the trend line of using large farm animals, Bio-Protein Technologies (Paris) has specialized in producing therapeutic monoclonals in rabbit milk. The main advantage of using rabbits is that their gestation time is only one month (as opposed to 9 months for goats and cows) and that the female rabbits mature sexually in just four months. Rabbits are also prolific breeders. These two features will more than compensate for the low milk production from each rabbit (0.25 litres of milk per day compared to 20 litres per day for the cow).

Expressing a human protein in animal milk has been the first step. The level of the therapeutic protein is hardly 1 per cent of the total proteins in the milk. The total proteins represent only 4 per cent of the milk by weight. Hence, the challenge will be to purify the monoclonals from milk proteins since a small but significant portion of the population has been allergic to milk proteins. The purification costs for processing transgenic milk will not be very different from those incurred in cell culture.

Safety Concern

The incidence of bovine spongiform encephalopathy (BSE), and transmissible spongiform encephalopathy (TSE) known as "**Madcow disease**" (in cows) and **scrapies** (in sheep) and difficult-to-detect pathogens such as prions prevalent in the beef and meat industry has raised a red flag on the safety of these products.

The transgenic firms counter that charge by pointing out that milk-producing animals do not typically harbour viral pathogens that plague humans. Besides, animal lineage can be traced rigorously. Once a pedigree is deemed clear of prions, the controlled environment in which the animals are reared guarantees safety. The regulatory agencies in the US and Europe are not overly concerned about the safety issues although they have set up appropriate guidelines.

Another issue that might come up in the future, concerns the possibility of the transgenic animals being slaughtered and brought into the food chain. The transgenic animals which express foreign protein in milk will also express it in other tissues at very low levels. Other concerns involve animal welfare. Calves and lambs produced through cloning often have higher birth weights and longer gestation times than regular ones. As a result, births are often difficult and require a caesarean delivery.

The Center for Science in the Public Interest, a watchdog organization in the US, emphasizes that transgenic animals are not released into the environment or allowed to enter the food supply without a thorough assessment by the government. Ultimately, it is not the actual safety issues or the declaration of safety by government agencies that will determine the success of the transgenic animal therapeutics, but the market economics and the public perception of the safety issues, as is the current situation with genetically modified crops.

HUMANIZED BOVINE IMMUNOGLOBULIN SYSTEM

Non-rodent transgenic animals, such as cows, chicken and rabbits, could be exploited in the biological production of human antibodies. In the September 2002 issue of *Nature Biology*, Kuroiwa *et al.*, a team of researchers from Japan and U.S., reported the generation of cloned calves that expressed human immunoglobins.

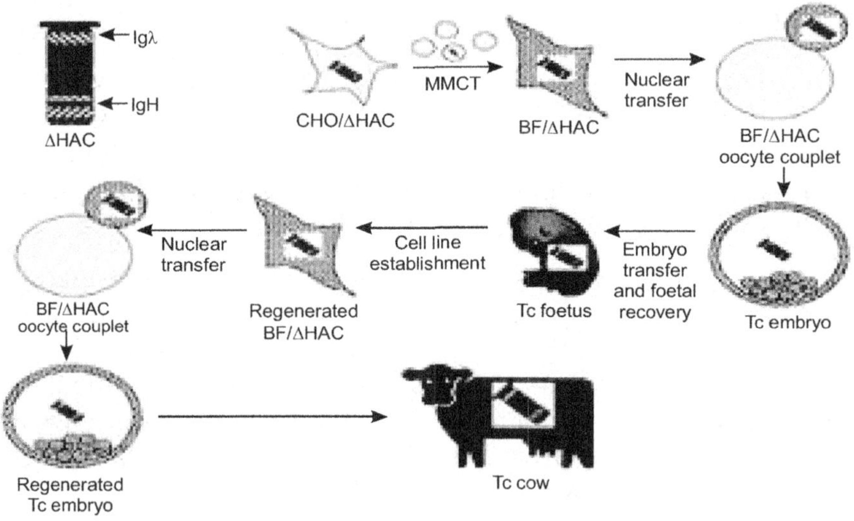

Figure 5.7 Production of cloned calf expressing human immunoglobulins (Kuroiwa, *et al.*, 2002)

A human artificial chromosome of approximately 10 mega bases in size was constructed to contain unrearranged human germ-line heavy-chain and λ-light-chain loci, and transferred into primary bovine fibroblasts, using microcell-mediated chromosome transfer (MMCT) procedure. This transformed fibroblast was used in a subsequent nuclear transfer step to produce homozygous heavy-chain knockout mutant calves. The resultant transgenic calves were named as **transchromosomic calves** (Figure 5.7).

The transgenic foetal cell lines were used in the second round of nuclear transfer. Six calves were generated, of which four were phenotypically normal and healthily survived. Analysis revealed that human artificial chromosome was stable, and heavy- and light-chain genes undergo functional rearrangement. The immunoglobin level was found to be in the range of 13–258 ng/ml blood. This system has been tremendously useful not only due to large quantities of antibodies but also the calves can be hyperimmunized with any antigen.

MONOCLONAL ANTIBODIES IN CHICKEN

"Transgenic chickens should be considered a near perfect pharmaceutical bioreactor for making large amounts of pure recombinant proteins" says Ann Gibbens, an avian biologist at the University of Guelph in Ontario, who did pioneering studies in the field and trained many of those scientists now achieving success. The idea of making drugs in chicken has been appealing. Each commercial hen lays about 250 plus eggs per year. Each egg contains nearly 4 grams of egg white comprising high proportion of the recombinant protein. Gibbons says that the final cost for purification of protein should be about $ 10 per gram; hundredfold less than the cost of current systems using mammalian cells. Moreover chicken flocks have been easy to ramp up in a month.

Encouraged by the published results in the April issue of *Nature Biotechnology*, 2002, AviGenics, a private company in Georgia, reporting the production of a marker enzyme, beta-lactamase, in egg whites of transgenic chickens, it has become interested in the production of therapeutic monoclonals from chickens. Another group in Ohio called OvImmune has also been involved in the production of antibodies from immunized chickens. OvImmune got in trouble with the FDA for circumventing conventional drug development protocols, offering its eggs to the community as a sort of health food.

Origen Therapeutics announced in August 29, 2005, at Burlingame, the first published scientific report of fully functional, human sequence monoclonals produced in chickens. The antibodies were expressed solely in the chicken oviduct and deposited into egg white in concentrations of 1–3 milligrams per egg. Moreover, antibodies produced in this manner demonstrated 10–100-fold greater cell-killing ability (ADCC) compared to therapeutic antibodies produced by conventional cell culture methods.

The new report has been published in the September issue of *Nature Biotechnology*, 2005, by researchers from Origen Therapeutics and their collaborators at Medarex, Texas; and A and M University and the University of California, Los Angeles. A research work briefly commenting on the potential impact of this development for the production of human therapeutic proteins has also been published in the September issue of *Nature Medicine,* 2005.

"This work demonstrates the potential for producing therapeutic proteins with enhanced properties in the eggs of chickens as an alternative to established mammalian cell culture systems," said Robert J. Etches, Origen Therapeutics vice-president, Research. "Antibodies produced by this method had very similar physical and biological characteristics to those produced in CHO cells, including nearly identical binding curves, similar affinities, and an equal ability to be internalized by antigen on prostate cancer cells. At the same time, chicken-produced antibodies lacked the sugar residue, fucose, which greatly increases their cell-killing activity compared to CHO-produced antibodies."

To create the antibody-producing chickens, the researchers first inserted into chicken the embryonic stem cells, the genes encoding the antibody, along with the regulatory sequences restricting its deposition to egg white. The stem cells were then introduced into chick embryos. At this stage of development, the embryonic stem cells can make significant contributions to the developing chicken (Figure 5.8). Resulting chimeras with large contributions from the stem cells laid eggs containing milligram amounts of antibody, which was then separated from the egg white proteins, generating the purified product. "Furthermore," Dr. Kay continued, "unlike other transgenic animal systems, the time from antibody identification to production in eggs can be as short as 8 months versus 18 months to 3 years for goats or cattle. The egg is sterile and stable, providing a good starting material for isolation and purification of the protein of interest.

Moreover, conditions for good manufacturing practices have long been established for vaccine production in chicken eggs."

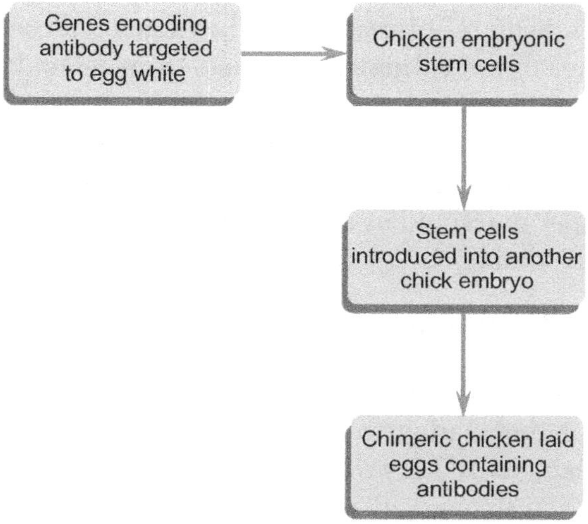
Genes encoding antibody targeted to egg white
Chicken embryonic stem cells
Stem cells introduced into another chick embryo
Chimeric chicken laid eggs containing antibodies

Figure 5.8 Flowchart showing the production of antibody-secreting chicken (*Nat. Biotech.*, 2005)

"We believe the chicken system is an attractive one for therapeutic protein production compared to either plant systems or to other transgenic animal systems," said Robert Kay, Origen Therapeutics President and Chief Executive Officer. "The fact that the chicken-produced anti-cancer antibodies show dramatically enhanced cell-killing activity, elevates the chicken system considerably relative to other non-traditional production technologies and some traditional cell culture methods as well." The introduction of this new chicken-based production technology will be of considerable interest to an industry coping with the commercial supply of an ever-increasing number of therapeutic antibodies." "This new technology has the potential to drive down the drug manufacturing costs, which could make medicines and health insurance plans less expensive for all of us." says Mathew E. Portnoy, Director of National Institute of General Medical Sciences at the National Institute of Health.

PROBLEMS INVOLVED IN TRANSGENIC ANIMALS

Time is a major problem in the production of transgenic animals. Though science has learned to make transgenic animals, it has not yet discovered methods to make a kid become a goat faster. Currently it takes 12 months to 3 years just to get the first batch of clinical product from a transgenic animal. Until the time problem is solved, the business model for transgenic animals has to be targeted on conversion of high-value, high-volume validated products from cell culture to transgenic production. This plan is best illustrated by Genzyme, which has the broadest partnering program, reporting over 60 molecules in various stages of development. Much of Genzyme's work is based on proteins that are either marketed or in late clinical stages with anticipation for high-volume demand.

A small herd of goats or sheep could replace a massive cell culture plant, not to mention the relative cost of grain and grass vs. culture media. However, one caution is that transgenic companies will need to work out the management of a large-production herd (over 150 animals), especially for proteins with high demands (>1 metric ton/year) that are expressed at the low range of transgenic volumes (<1 gram per litre of milk).

REFERENCES

Alt, F.W., Blackwell, T.K. and Yancopoulos, G.D. (1985). "Immunoglobulin genes in transgenic mice." *Trends Genet.* **1**: 231–236.

Bruggemann, M., Caskey, H.M., Teale, C., Waldmann, H., Williams, G.T., Surani, M.A. and Neuberger, M.S. (1989). "A repertoire of monoclonal antibodies with human heavy chains from transgenic mice." *Proc. Natl. Acad. Sci.* USA. **86**: 6709–6713.

Fernandez, M., Crawford, L. and Hefferan, C. (2002). Pharming the field: A look at the benefits and risks of bioengineering plants to produce pharmaceuticals. Proceedings from a workshop sponsored by the Pew Initiative on Food and Biotechnology, FDA, and USDA.

Ferrante, E., Simpson, D. and Timothy, C. Scott. (2001). "A review of the progression of transgenic plants used to produce plantibodies for human usage." *Biol.* & *Biomed. Sci. Issue.*

Fiedler, U. and Conrad, U. (1995). "High-level production and long-term storage of engineered antibodies in transgenic tobacco seeds." *Biol. Technology.* **13**: 1090–1093.

Fishwild, D. *et al.* (1996). "High-avidity human IgG kappa monoclonal antibodies from a novel strain of minilocus transgenic mice." *Nat. Biotechnol.* **14**: 845–851.

Green, L.L. and Jakobovits, A. (1998). "Regulation of B cell development by variable gene complexity in mice reconstituted with human immunoglobulin yeast artificial chromosomes." *J. Exp. Med.* **188**: 483–495.

Green, L.L. *et al.* (1994). "Antigen-specific human monoclonal antibodies from mice engineered with human Ig heavy and light chain YACs." *Nat. Genet.* **7**: 13–21.

Haitt, A., Cafferkey, R. and Bowdish, K. (1989). "Production of antibodies in transgenic plants." *Nature.* **342**: 76–78.

Hiatt, A. (1990). "Antibodies produced in plants." *Nature.* **344**: 419–470.

Hiatt, A. and Ma, J.K.C. (1992). "Monoclonal antibody engineering in plants." *FEBS Lett.* **307**: 71–7.

Hiatt, A. and Ma, J.K.C. (1993). "Characterization and applications of antibodies produced in plants." *Int. Rev. Immunol.* **10**: 139–152.

Ishida, I., Tomizuka, K., Yoshida, H., Tahara, T., Takahashi, N., Ohguma, A., Tanaka, S., Umehashi, M., Maeda, H., Nozaki, C., Halk, E. and Lonberg, N. (2002). "Production of human monoclonal and polyclonal antibodies in transchromosomic animals." *Cloning Stem Cells.* **4**: 91–102.

Julian, K.C., Hiatt, A., Hein, M. *et al.* (1995). "Generation and assembly of secretory antibodies in plants." *Science.* **268:** (5211), 716–719.

Julian, K.C., Lehner, T., Stabila, P. *et al.* (1994). "Assembly of monoclonal antibodies with IgG1 and IgA heavy chain domains in transgenic tobacco plants." *Eur. J. Immunol.* **24**: 131–138.

Kuroiwa, Y., Kasinathan, P., Choi, J.Y., Naeem, R., Tomizuka, K., Sullivan, J.E., Knott, G.J., Duteau, A., Goldsby, R.A.,Osborne, A.B., Ishida, I. and Robl, M.J. (2002). "Cloned transchromosomic calves producing human immunoglobulin." *Nat. Biotechnol.* **20**: 889–894.

Larrick, J.W., Yu, L., Chen, J. *et al.* (1998). "Production of antibodies in transgenic plants." *Research in Immunology.* **149**(6): 603–8.

Little, M., Kipriyanov, S.M., Le Gall, F. and Moldenhauer, G. (2000). "Of mice and men: hybridoma and recombinant antibodies." *Immunol. Today.* Vol. 21. No. 8, 364.

Ma, J.K. and Hein, M.B.(1995). "Plant antibodies for immunotherapy." *Plant Physiol.* **109**: 341–346.

Ma, J.K. and Hein, M.B. (1995). "Immunotherapeutic potential of antibodies produced in plants." *Trends Biotechnol.* **13**: 522–527.

Ma, J.K., Hikmat, B.Y., Wycoff, K., Vine, N.D., Chargelegue, D., Yu, L. *et al.* (1998). "Characterization of a recombinant plant monoclonal secretory antibody and preventive immunotherapy in humans." *Nat. Med.* **4**: 601–606.

Ma, J.K., Hiatt, A., Hein, M., Vine, N., Wang, F., Stabila, P. *et al.* (1995). "Generation and assembly of Secretory antibodies in plants." *Science.* **268**: 716–719.

Miele, L. (1997). "Plants as bioreactors for biopharmaceuticals: regulatory considerations." *Trends Biotechnol.* **15**: 45–50.

Nils Lonberg *et al.* (1994). "Antigen-specific human antibodies from mice comprising four distinct genetic modifications." *Nature.* **368**: 856–859.

Nils Lonberg (2005). "Human antibodies from transgenic animals." *Nat. Biotechnol.* **23**: 1117–1125.

Parkinson, S. (1995). "Production of monoclonal antibodies in the milk of transgenic animals." Exploiting Transgenic Technology, IBC Symposium, November, San Diego.

Renner, C., Hartman, F. and Pfreundschuh, M. (2001). "The future of monoclonal antibody engineering." *Annals of Haematology.* **80**: (3B) 127–129.

Schade and Rudiger, *et al.* (1996). "The production of Avian (egg yolk) antibodies Igγ, ATLA (Alternative to laboratory animals)." Vol. **24**: 925–934.

Sharma, A.K., Mohanty, A., Singh, Y. and Tyagi, A.K. (1999). "Transgenic plants for the production of edible vaccines and antibodies for immunotherapy." www.ias.ac.in/currsci/aug25/articles20.htm.

Stoger, E., Vaquero, C., Torres, E., Sack, M., Nicholson, L., Drossard, J., Williams, S., Keen, D., Perrin, Y., Christou, P. and Fisher, R. (2000). "Cereal crops as viable production and storage systems for pharmaceutical scFv antibodies." *Plant Mol. Biol.* **42**: 583–590.

Talwar, G.P. (2001). Emerging era of very safe and novel therapeutics with humanized recombinant antibodies produced in plants: The 15th National Conference on inhouse R & D in Industry, November 22–23, New Delhi.

Tomizuka, K., Tomizuka, T., Shinohara, H., Yoshida, H., Uejima, A., Ohguma, S., Tanaka, K., Oshimura, M.S. and Ishida, I. (2000). "Double trans-chromosomic mice: maintenance of two individual human chromosome fragments containing Ig heavy and kappa loci and expression of fully human antibodies." *Proc. Natl. Acad. Sci.* USA. **97**: 722–727.

Whitelam, G.C., Cockburn, W., Owen, M.R.L. (1994). "Antibody production in transgenic plants." *Biochemical Society Transactions.* **22**: 940–943.

Wycoff, K.L. (2004). "Secretory IgA antibodies from plants." *Current Pharmceu. Design.* 10.

6

APPLICATIONS OF MONOCLONALS

Many of the early monoclonals produced were used in viral research laboratories to study various viral antigenic determinants, to characterize viruses, to isolate viral components for biochemical studies, to find out the antigenic relationship among the individual determinants of viruses and also to identify variant viruses that were previously unrecognizable.

Due to its enormous potential and remarkable specificity, monoclonal technology has moved from viral research laboratories to clinical or commercial applications. The highly purified monoclonals are now widely used in commercial diagnostic kits, in histocompatibility testing and in diagnosis and treatment of multitude of diseases including diabetes. As a consequence, a new industry has grown up and in recent years several biotechnology companies have been started, whose income is mainly from the sales of monoclonals.

Monoclonals were one among the first handful of approved therapeutic molecules g enerated by modern biotechnology. In 1986, the first murine monoclonal approved for treating acute organ transplant rejection was muromonab-CD3 (Orthoclone OKT3, Ortho Biotech, Bridgewater, NJ, USA), targeting the CD3 antigen of T cells. However, after the introduction of this monoclonal, 8 years elapsed before the next therapeutic monoclonal was approved by the US Food and Drug Administration (FDA). An important factor that contributed to this gap was the observed immunogenicity of mouse antibodies in human patients, which can lead to rapid clearance, reduced efficacy and an increased risk of infusion reactions. This immunogenicity ranges from relatively benign fevers and rashes to severe cardiopulmonary and anaphylactic-like adverse events.

Some applications of monoclonals are discussed in the following sections.

IMMUNOASSAY

Immunoassay technology has been virtually overhauled by the introduction of monoclonals. The impact of the monoclonal revolution can be realized by the sheer rapidity of incorporation of monoclonal reagents into immunoassay. Today nearly 70% of the immunoassay for various analytes uses monoclonal antibodies. The sensitivity and reproducibility of the existing immunoassays for blood-group antigens, serum antigens, MHC antigens, fibronectin, interleukins, complement components, interferons, progesterone, gastrin, plasminogen, oestrogen and human gonadotrophin have been improved with the use of monoclonals.

Monoclonals have found widespread use in laboratories of all types, and in various applications from drug discovery to detecting drugs of abuse in athletes and horses. The routinely used conventional immunoassays like agglutination, RIA, IRMA, ELISA, EIA, EMIT, and FIA are now replaced by FACS (fluorescent-activated cell sorter), time-resolved fluoroimmunoassay, liposome-bound immunoassay, chemiluminescent, ELISPOT, metallo-immunoassay, immunocapture (using abzymes), etc. Many of these later immunoassays have developed a high degree of automation with specialized instrumentation. Visible immunoassay with monoclonals conjugated directly or indirectly with fine-coloured particles, is the more recent innovation. These assays are very sensitive and so even nanogram amounts of the analyte can be detected within few hours or even within minutes.

Agglutination

After Karl Landsteiner's groundbreaking discovery of blood groups, blood-typing tests have been devised. For more than seven decades, agglutination was tested only with polyclonal antisera. But when the monoclonal antibody era was born, blood-typing laboratories started using a full range of high-quality monoclonal-derived blood grouping reagents manufactured to the highest standards.

Radioimmunoassay

RIA is a technique first developed in 1960. It is used for measuring the antigen (or antibody) that involves competitive binding of the radiolabelled monoclonal antibody (or antigen). It is a long, tedious and costly technique to be carried out. It has to be performed by trained and licensed persons and requires sophisticated radioactive counter for measurement. High sensitivity, high specificity and low interference in samples of patients with multiple therapy, simple handling of a great number of samples per person and good interlaboratory reproducibility are the advantages of this technique.

Immunoradiometric Assay

IRMA is a single-site assay in which excess of labelled antigens are added to the sample containing antibody, and the immunocomplex formed is allowed to precipitate and after settling, the solid and liquid phases are quantified for radioactive counts. In the two-site IRMA, immobilized monoclonals are used.

ELISA

Enzyme-labelled immunosorbent assay is used to quantify the antibody using an enzyme-labelled ligand of antibody. The coloured product formed after the addition of colourless substrate is directly propotional to the amount of antibody in the sample. The enzymes commonly used are horseradish peroxidase and alkaline phosphatase and the ligand commonly used is protein-A of *S. aureus* or a secondary antibody. Table 6.1 gives the list of colourless substrates used and the colour of the products formed by the two enzymes used for labelling. It is a simple, cheap, eco-friendly assay performed within a few hours. It includes repeated washing procedures to wash off the unbound excess reagents (Figure 6.1).

Table 6.1 Substrates used for labelled enzymes and the colour of the products obtained

Enzymes	Substrates	Colour
Horseradish peroxidase	Diaminobenzidine	Brown
	Tetramethyl benzidine	Blue
	Aminoethyl carbazole	Red
	4-chloro-1-naphthol	Blue-black
Alkaline phosphatase	Naphthol AS phosphate + fast red, blue or violet	Red, blue or violet
	Bromochloroindolyl phosphate + nitroblue tetrazolium	Blue

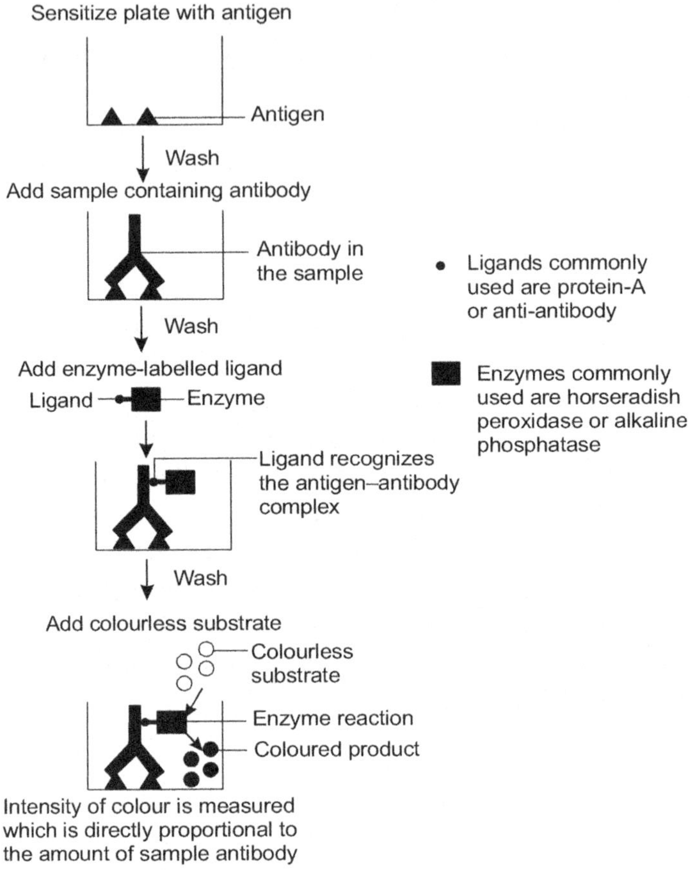
Sensitize plate with antigen
Antigen
Wash
Add sample containing antibody
Antibody in the sample
Wash
Add enzyme-labelled ligand
Ligand
Enzyme
Ligand recognizes the antigen–antibody complex
Wash
Add colourless substrate
Colourless substrate
Enzyme reaction
Coloured product
Intensity of colour is measured which is directly proportional to the amount of sample antibody
• Ligands commonly used are protein-A or anti-antibody
■ Enzymes commonly used are horseradish peroxidase or alkaline phosphatase

Figure 6.1 Diagram showing the principle of ELISA

The sensitivity of this assay is increased by several ways. One way is to amplify the signal of one enzyme by the use of another enzyme. The product (NAD) of the first enzyme (alkaline phosphatase) is used to set up a cycle in which the second enzyme (diaphorase) generates large amount of coloured products (tetrazolium violet). This cycle is kept by alcohol dehydrogenase, that simply helps in amplification and production of NADH. Hundredfold higher sensitivity is achieved with one molecule of NAD generated by the enzyme alkaline phosphatase. It reacts with the colourless formazan (substrate) in the presence of diaphorase (second enzyme) and produces as many as 100 tetrazolium violet molecules (Figure 6.2).

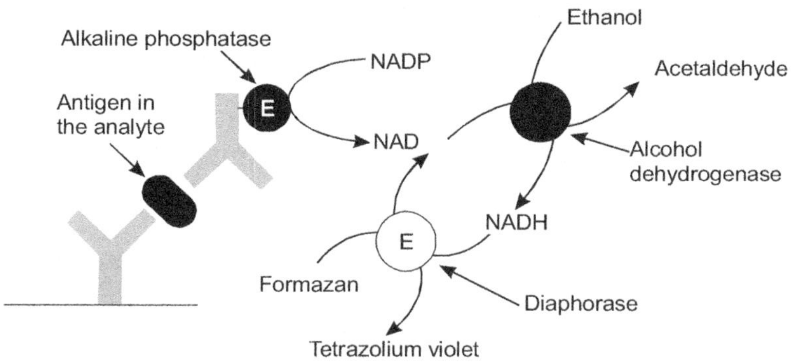

Figure 6.2 Amplification mechanism for enhancing the signal in ELISA (King, D.J., 1998)

Avidin–biotin ELISA The most useful strategy to date has been the use of avidin–biotin or biotin–streptavidin interaction. Biotinylated antibodies react with streptavidin that is labelled with enzymes or fluorescent substances. For instance, thyroglobin molecules labelled with up to 480 europium chelate groups are attached to streptavidin and are used to detect the biotinylated antibody resulting in an amplification of 4500–6750-fold. Biotinylated-Fab fusion proteins can be produced by recombinant DNA technology.

EMIT Enzyme multiplied immunotechnique is a homogeneous assay, in which there is no need to separate the bound and free antibody. It is a simple, easy-to-use assay, although sensitivity is often relatively low. A later version of this system uses a recombinant form of the enzyme which is more suitable for automation. EMIT can be performed for drugs of abuse like morphine, barbiturates and amphetamine, for cardiovascular drugs like digoxin, propranolol and quinidine, for antidepressants like amitryptylin, nortryptylin and imipramine, for chemotherapeutics like methotrexate and gentamicin and for hormones like T3, T4, oestriol and cortisol.

ELISPOT This assay is used to detect the secretions of cells, like cytokines, which are otherwise difficult to detect due to the short period of secretion and small amount. It is performed in a 96-well microtitre plate. In the first step, the wells are coated with high-affinity monoclonals to the cytokines to be investigated. In the second step, cytokine-secreting cells are added and incubated for 6 to 48 hours. During this time the cytokines are captured by monoclonals.

After washing the cells, biotinylated antibodies raised against the second epitope of cytokine is added. Streptavidin-conjugated enzyme is then added and tested with a colourless substrate and incubated. The appearance of coloured spots can be counted or analysed with an image analyser. The number of spots is compared with the number of cells added (Figure 6.3).

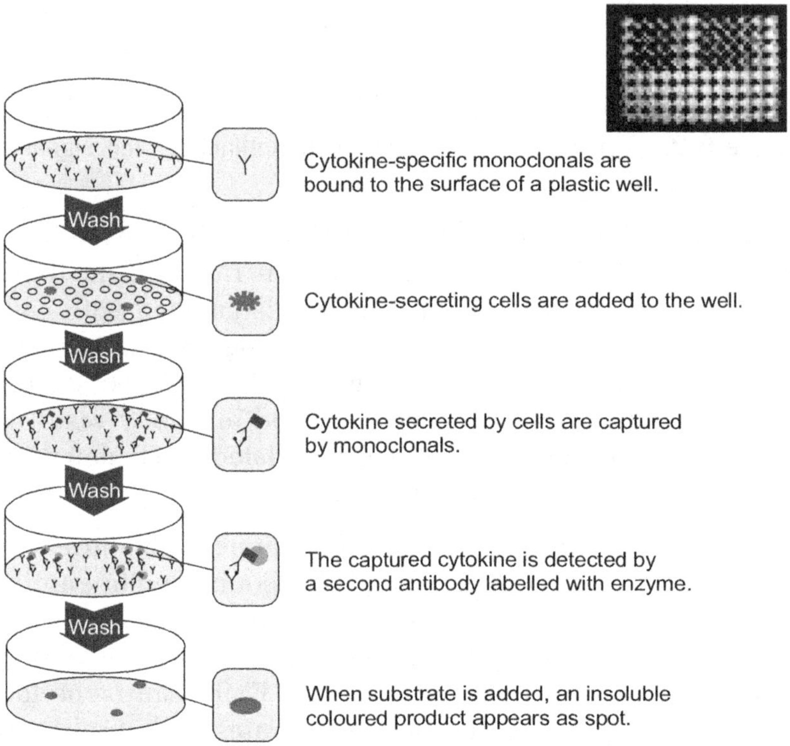

Figure 6.3 Use of monoclonals in ELISPOT (www.mabtech.se/images/elispot_method.gif)

Chemiluminescent assay Some enzymatic reactions produce light, and this can be measured to detect the product formation. In the chemiluminescent reaction, the product is in an electrically excited state, emitting light. This assay is extremely sensitive, since the light produced can be captured by a photographic film over days or weeks, but is hard to quantify, because not all the light released by a reaction will be detected. For example, the detection of horseradish peroxidase by enzymatic chemiluminescence (ECL) is a common method of detecting antibodies in western blotting. The secondary antibody is labelled with horseradish perodixase. The gel with bound antibody is incubated with luminol and hydrogen peroxide. The light produced is detected by a photographic film.

$$\text{Luminol} + H_2O_2 \xrightarrow[\text{Labelled antibody}]{\substack{\text{Horseradish}\\ \text{peroxidase}}} \text{3-Amino naphthalate} + \text{Light}$$

Another example is the detection of antibodies tagged with the enzyme luciferase present in fireflies, which produces light from its substrate luciferin.

$$\text{Luciferin} + \text{ATP} + O_2 \xrightarrow{\text{Luciferase}} \text{Oxyluciferin} + \text{Light}$$

Most recent and useful chemiluminescent labels are acridium esters, and aequorin, a calcium-activated photoprotein, isolated from the jelly fish *Aequorea victoria.* Aequorin-conjugated monoclonal antibodies have been used for the development of sensitive assays for hormones like thyrotrophin, chorionic gonadotrophin, lutrophin (LH) and follitrophin (CFSH). Fab-aequorin recombinant fusion proteins are available for this assay. Chemiluminescent reagents show 5000 times more amplification.

Immunoliposome assay Another approach to amplify signal is to attach the antibody of interest to a liposome, into which are entrapped many detectable molecules such as enzymes or fluorescent or chemiluminescent reagents. The liposomes are lysed at the end of the assays and the contents are quantitated. Recombinant lipid-tagged-scFv fragments obtained from *E. coli* can be incorporated into liposomes and used with high efficiency.

CEDIA It is a cloned enzyme donor immunoassay. In this assay, two inactive fragments of β-galactosidase—one an enzyme donor (ED) and the other an enzyme acceptor (EA)—are produced by recombinant means which when mixed together associate to produce active enzymes. When the enzyme donor (ED) is conjugated to antigen, monoclonal antibody binding prevents association and formation of the active enzyme. If no analyte is present in the sample, the ED-A complex binds to antibody and is not available to interact with the EA to form a functional enzyme. If analyte is present in the sample, it competes with the ED-A complex for the antibody-binding sites. This frees the ED-A complex to interact with EA (Figure 6.4).

Such homogeneous systems have been developed for assaying many low-molecular-weight substances such as cortisol, ferritin, digoxin, folate, vitamin B_{12} and drugs of abuse.

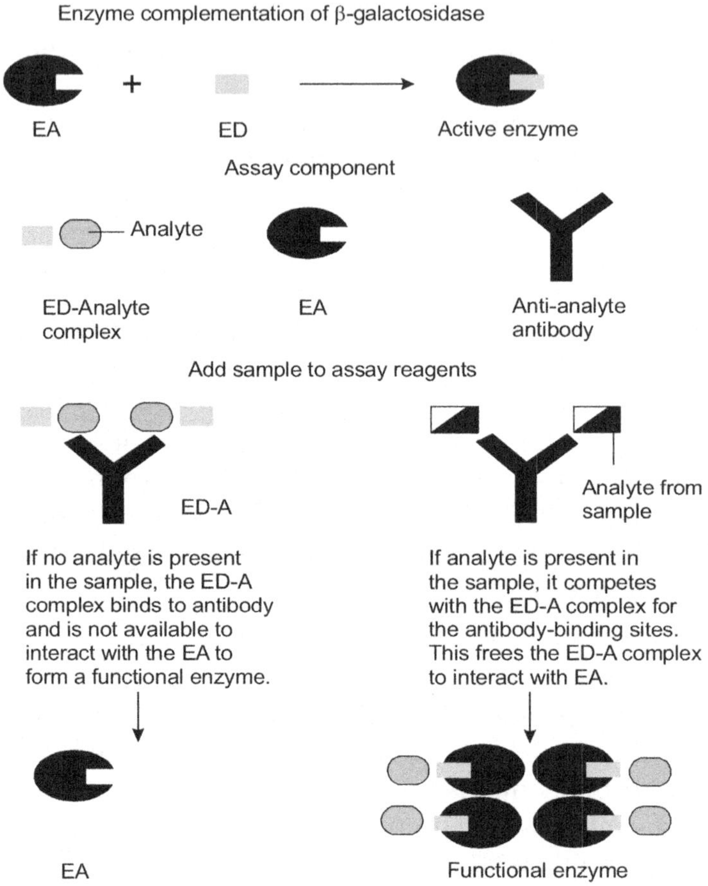

Figure 6.4 Diagram showing the principle of CEDIA

Immuno-PCR

Recently, a sensitive detection system for immunoassay, termed immuno-PCR utilizes the monoclonal antibody coupled to DNA through the avidin–biotin system. First the streptavidin is bound to an immobilized monoclonal through a protein A–streptavidin conjugate. This is then used to capture the biotinylated DNA which is then detected and amplified through PCR with the resulting products analysed by electrophoresis.

With the development of novel monoclonals, new antigens have been discovered and assays for these analytes have been developed. The newly developed monoclonals have been used to type the bacterial or viral strains and to detect the fungal or protozoan pathogens. But a major disadvantage of using monoclonals in viral typing is the antigenic shift or drift that may lead to changes in the antigenic proteins.

IMMUNOHISTOCHEMISTRY

One of the widely used applications of monoclonals is in immunohistochemistry. Since the introduction of monoclonals, immunohistochemistry has become an important tool in research, surgical pathology and oncology to detect and locate the antigens within the frozen or fixed cells or tissues. Fluorescent-labelled monoclonals are used in immunohistochemistry to identify the distribution of antigens in tissue sections, to determine cell types, to detect cellular infiltrates, to determine their active state, to localize adhesive molecules, to identify locally produced cytokines, to quantify apoptosis, to detect pathogens, to monitor disease progression in tissue biopsy and to identify the over-expression of antigens on solid tumours.

Monoclonals are used in FACS to separate different cell types from a mixture. The cells in the sample mixture are labelled with different fluorescent substances and are allowed to pass through a fine capillary tube to form a drop, each containing a single cell. They are excited by a laser beam and fluorescent cells can be separated according to their charge, granularity, and the secretion in electrostatic charger, which deflect them into different containers. FACS allows the separation of immune cells with different markers, different interleukin secretions, etc. (Figure 6.5).

High homogeneity, absence of non-specificity, ease of characterization and no batch-to-batch or lot-to-lot variability, make monoclonals more advantageous than their polyclonal counterparts. However, there are certain pitfalls in the use of monoclonals in immunohistochemistry. The frozen tissues or formalin-treated tissues for target epitope must survive fixation. Cross-reactivity of monoclonals cannot be removed. For optimum performance, monoclonals need maintenance of appropriate pH. In frozen sections, sometimes well-characterized antigens cannot be detected whereas a variety of hidden antigens can be detected. In such cases, the sensitivity of this method can be increased by pretreating the fixed tissues in various ways.

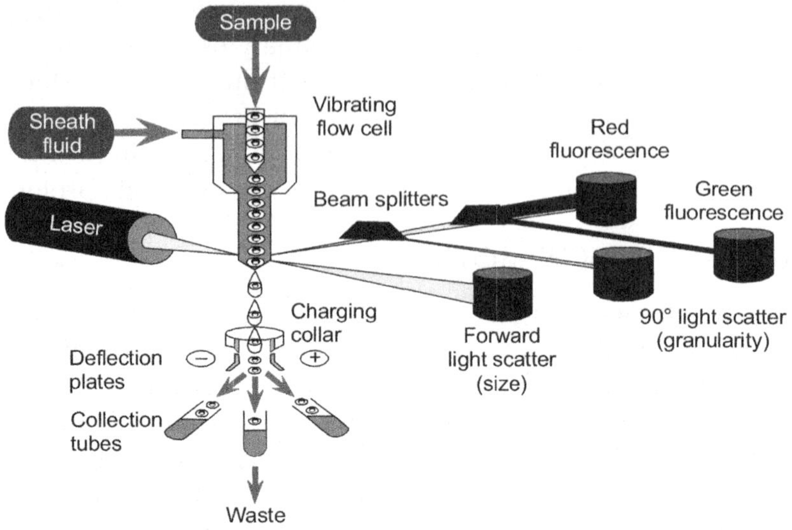

Figure 6.5 Fluorescent associated cell sorter (Roitt, *et al.*, 2002)

A major problem with the use of fluorescent labels, however, is the background level of fluorescence generated by many biological substances and by the plastics used for immobilization. The background fluorescence is due to very rapid events, with a fluorescence lifetime of 100 nanoseconds or less. Hence, conventional fluorescent compounds used in immunohistochemistry such as fluorescein that emit fluorescence for a short period, are of relatively little use. This can be overcome by the use of time-resolved fluorescent technique. In this technique, lanthanide chelates are used as the fluorescent reagents. It emits light for longer periods (1000 microseconds or more). When pulses of light are used to excite the fluorescent material present, after a short time interval of a few hundred microseconds (during which the background fluorescent interference fades away), the light emitted by lanthanide chelates is observed.

WESTERN BLOTTING (IMMUNOBLOTTING)

This technique is used to detect the protein antigens that are first separated by SDS-PAGE. The separated proteins are transferred to nitrocellulose or polyvinylidene difluoride. The antigens are then detected by a suitable specific primary antibody, which is later detected by a general secondary labelled antibody. The labels commonly used are either ^{125}I followed by exposure to either X-rays, or to colloidal gold with silver enhancement, or to enzymes with coloured substances

(Figure 6.6). Even though both monoclonals and polyclonals are used in western blotting, monoclonals offer excellent specificity.

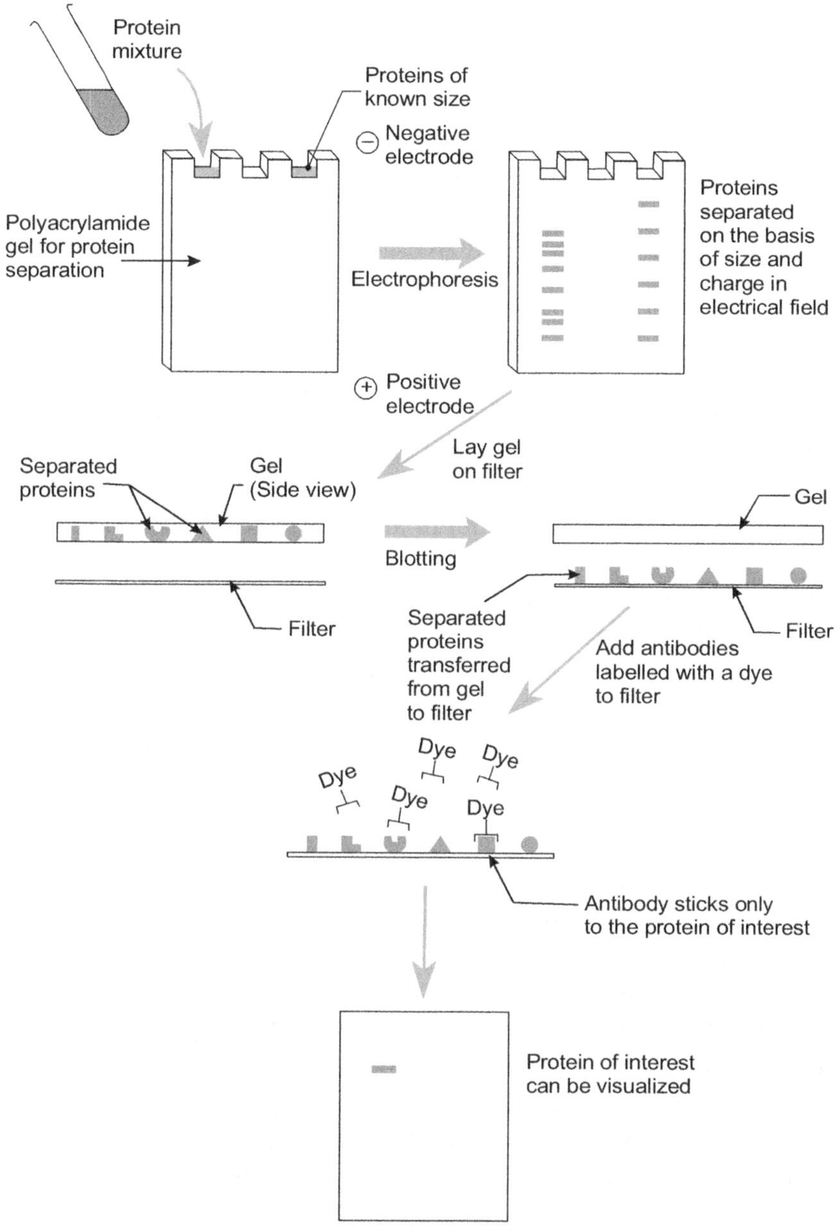

Figure 6.6 Diagrammatic representation of western blotting (http://www.bseinquiry.gov.uk/report/volume2/images/chaptec8.gif)

This technique is extensively used in research for the detection and analysis of traces of proteins, expression of recombinant proteins and contaminant proteins, etc. Monoclonals allow the detection of some proteins which would not be possible otherwise. For example, the degradative role of synovial matrix metalloprotease enzymes in human arthritis was analysed by observing the cleaved product of aggrecan, a major proteoglycan in cartilage, which bears the weight of the body. This technique is also used to detect HIV types and subtypes and to measure carbohydrate-deficient transferrins in alcohol abuse. It is also used to isolate recombinant fusion proteins expressed in *E. coli* in a single step.

DIAGNOSIS OF PREGNANCY

The diagnostic applications of monoclonals are by far the most advanced, especially for tests that are performed on body fluids such as blood and urine samples. Nearly one-third of the 150 diagnostic monoclonals that have been approved by U.S. Food and Drug Administration are for pregnancy detection. The urine of pregnant women contains a hormone called human chorionic gonadotrophin (hCG), which is secreted by the placenta. It is an oncofoetal protein made soon after conception and has a critical role in implantation of the embryo on to the womb. The use of monoclonal to detect this hormone now permits the diagnosis of pregnancy as early as a week or two after conception. Thus pregnancy can be tested at home within three minutes with 99% sensitivity and specificity.

The commercial (hCG) test strip (Figure 6.7) includes immunochromatographic strips (or membrane) with three zones of antibody. The first zone has mouse monoclonal antibodies labelled with enzyme. The second zone is the test zone. It has polyclonal antibodies and a dye-substrate fixed or immobilized. The third zone is a control zone. It has anti-mouse antibodies and a dye-substrate. It has two windows, one small and another large. The small window or sample window is used to apply urine sample. The large window or test window is used to show both the control and test lines.

When the urine with hCG is applied on the sample window, it reaches the first zone consisting of mouse monoclonals labelled with enzymes. Some of the enzyme-linked antibody molecules attach with hCG.

The urine detaches the bound and unbound antibodies and continues to migrate and reaches the second zone. In the second zone, hCG and enzyme-bound antibodies bind to the immobilized antibodies and the enzyme reacts with the dye and produces red colour in the test zone. The unbound antibodies move along with the membrane, reach the control zone and combine with the antimouse monoclonals, and produce the red colour in the control zone.

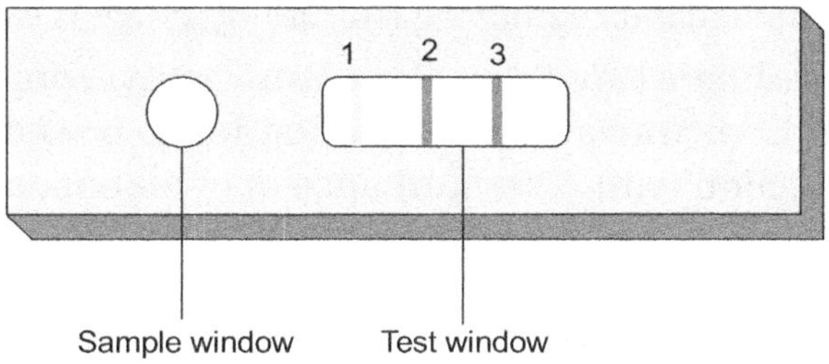

1—Zone containing mouse monoclonals labelled with enzyme;
2—Test zone (T) containing immobilized polyclonal with dye-substrate
3—Control zone (C) containing anti-mouse antibody with dye-substrate

Figure 6.7 Diagram showing pregnancy strip

One of the following results will be observed in the strip.

Positive Appearance of two distinct red lines—one at "T" (test) and the other at "C" (control).

Negative A single red band at the "C" region, indicating that a detectable level of hCG is not present.

Invalid If no band appears at the "C", or incomplete or "beaded" bands appear at either the test region or the control region, the test is invalid and should be repeated.

Many variants of such tests are now available, the use of which are also extended to ovulation prediction (luteinizing hormone, follicular stimulating hormone), and hormone oestrone-3-glucurinide. Similar tests are under development or in use for applications such as identification of the viruses such as HIV, HBV, etc., drugs of abuse and screening and monitoring of the disease markers.

PREVENTION OF PREGNANCY

Anti-hCG monoclonals inactivate the hCG and block the onset of pregnancy as well as its sustenance over the first 7 weeks. Thus these can be used

- as emergency contraceptive to prevent the onset of pregnancy, with the advantage of use up to 6–9 days after unprotected sex, whereas the Yuzpes regime is effective only up to 48–72 hours.
- in menstrual regulation (inducing delayed menstruation).
- or as home leave, vacation contraceptive, providing protection from an unwanted pregnancy for 4–6 weeks after a single injection without disturbance of ovulation, spotting, bleeding irregularities or mood change.

DIAGNOSIS OF DISEASES

Monoclonals provide a valuable diagnostic tool for health practitioners. In the immunodiagnosis of sexually transmitted diseases and of a number of viral, bacterial, protozoan and helminthic diseases, monoclonals form the major component. Monoclonals allow a rapid detection of hepatitis B and C, influenza, HSV 1 and 2, respiratory syncytial virus (RSV), rabies, cytomegalovirus (CMV), HIV, chlamydial and streptococcal infections. The diagnosis is also being improved by the availability of new FDA-approved monoclonals. Once the causative agent is identified, the method of treatment can be decided and the severity of the disease is reduced.

For example, gonorrhoeal and chlamydial infections in genitalia cause similar symptoms, but need a different treatment. To identify these infections, it takes at least about 3 to 7 days. Robert Nowinski and his colleagues at Genetic Systems Corporation and the University of Washington School of Medicine have produced monoclonals that can identify both the infections in as little as 15–20 minutes, thereby allowing rapid treatment. Similarly monoclonals against HIV1 and HIV2 can diagnose and determine the viral type within 15–20 minutes, in contrast to the earlier diagnostic tests which take 3 to 6 days. Monoclonals have been employed for immunocharacterization and immunodetection of intracellular helminthic infections like *Trichinella spiralis, Trypanosoma cruzi* and *Fasciola hepatica* which are very difficult to detect using earlier techniques.

Table 6.2 Some immunodiagnostic monoclonal kits

Polypeptide	Infectious disease	Tumour markers	Cytokines	Drug monitoring	Miscellaneous targets
Chorionic gonado trophin	Chlamydia	CEA	Interleukin 1–8	Theophylline	Vitamin B_{12}
Growth hormone, TSH, LH, FSH	Herpes	PSA	Colony stimulating factor	Gentamicin	Ferritin
Prolactin	Rubella, Legionella	IL-2 receptor	TNF EGF	Cyclosporin	Fibrin degradation product

Many monoclonals have been developed for use as immunodiagnostic agents for a variety of different compounds and pathogenic organisms (Table 6.2). Many of these diagnostics utilize strips of paper impregnated with an appropriate monoclonal antibody. Since the diagnostic products are relatively inexpensive to produce and are projected to have growing market, many biotechnology companies have entered this field.

TUMOUR DETECTION AND IMAGING

In *in vivo* imaging, monoclonals specific for certain tumour-associated antigens are tagged with radioactive materials and are injected into the bloodstream or elsewhere in the body. They bind to cancer cells thereby concentrating the radioactivity at the tumour site, indicating its location during imaging by X-rays. Monoclonal imaging technology can also detect the tumour-shed antigens in the blood and cancer metastases to nearby lymph nodes, which would be undetected by other scanning techniques (Figures 6.8 and 6.9). Monoclonals, coupled to gamma-emitting radioisotopes, have also been used to image tumours with a gamma camera for the purpose of diagnosing and monitoring the tumour spread.

Monoclonals can also be labelled with positron emission isotopes for use in positron emission tomography scanning. This technique is known as radioimmunodetection (RID) or radioimmunoscintigraphy (RIS). Accuracy in detection is improved by advances in instrumentation as single photon emission computerized tomography (SPECT). An alternative form of radioimmunodetection is used to aid surgical procedures, known as radioimmuno-guided surgery (RIGS).

In this technique, radiolabelled monoclonals localized to the site of disease are detected by a hand-held monitor to allow simple identification and resection of diseased tissue during surgery.

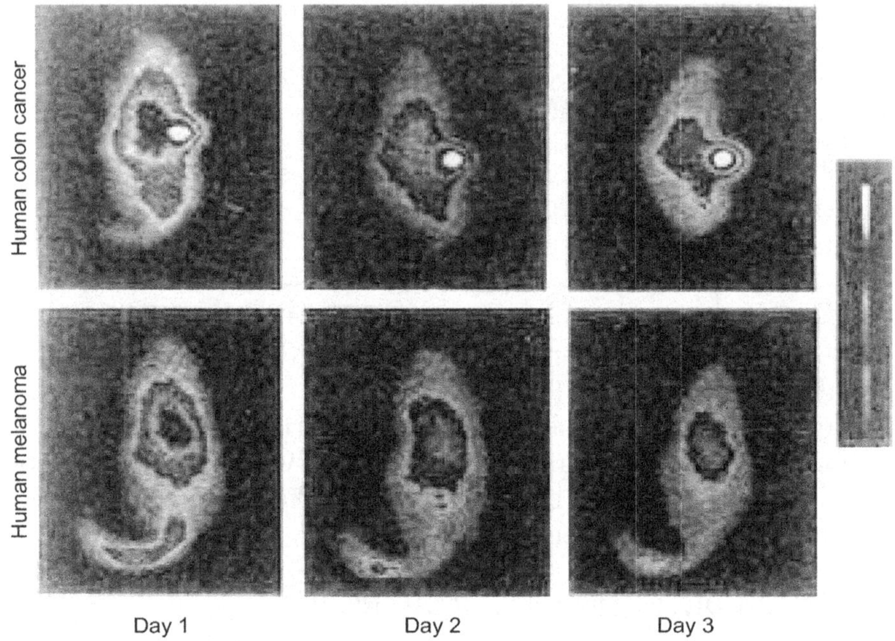

Figure 6.8 Use of monoclonals to locate cancer cells that have been transplanted into mice. The antibody labelled with radioactive iodine recognizes the antigen present on colon cells but not on myeloma cells. Radioactivity is detected by the first day of injection and persists till the third day, when radioactivity is less in blood. The antibody does not concentrate in the melanoma(Marx, J.L., 1989).

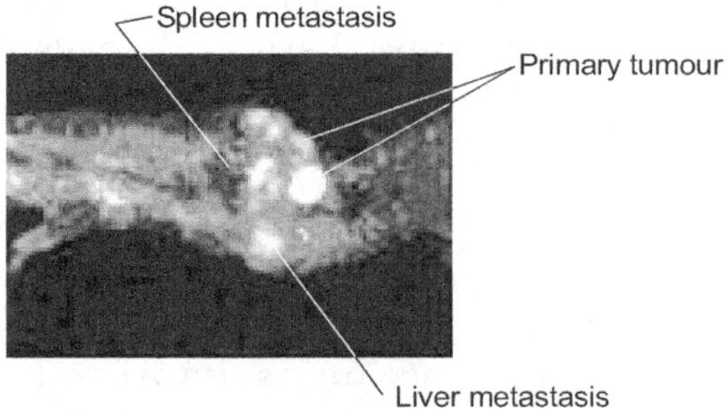

Figure 6.9 Whole-body image of the pancreatic carcinoma growing on pancreas with liver and spleen metastasis

Although the monoclonal imaging technology has promised in detecting and localizing cancers and metastases, there are a number of obstacles to the widespread use of monoclonals. All cancers do not share common tumour-specific antigens. Antigens should not be shed from the tumour into the blood. High-quality tumour imaging can be achieved through optimization of the immunoconjugate and through the use of alternative strategies to overcome the individual problems of a particular antigen. Table 6.3 shows the list of FDA-approved commercially marketed monoclonals for the diagnosis of cancer.

Table 6.3 List of FDA-approved monoclonals used for detection of different cancers

Product	Therapeutic indications	Year
CEA-Scan (Arcitumomab)	Recurrent and metastatic colorectal cancer	1996
Myoscint (Imciromab)	Myocardial infarction imaging agent	1996
Oncoscint CR/OV (Satumomab pendetide)	Detection/staging/follow-up of colorectal and ovarian cancer	1996
Prostoscint PSMA (Capromab pendetide)	Detection/staging/follow-up of prostate adenocarcinoma	1996
Technomab k1 (Fab)-HMW-MAA (Murine monoclonal)	Diagnosis of cutaneous melanoma lesions	1996
	Detection of small cell lung cancer	1996
Indimacis 125 (Igovamab) (Murine monoclonal)	Diagnosis of ovarian cancer	1996
Verluma (Nofetumomab)	Diagnosis of small cell lung cancer	1996
Leukoscan (Sulesomab) (Fab)NCA	Diagnostic imaging for infection and inflammation of patients with osteomyelitis	1997
Humaspect (Votumumab) Cytokinitin TSA	Detection of carcinoma of colon and rectum	1998

The specificity, density and shedding of tumour-associated antigens; the specificity, molecular size, affinity and avidity of antibody; the method of attachment, half-life and emission energy of radioactive substances; and location in the body, size and vascularization of tumour are several factors that affect radioimmunodetection of cancer. The common radioisotopes used for tumour imaging are ^{131}I, ^{99}mTc (technetium), metallic radionuclide, Indium-111 and gallium-67. Positron emission tomography (PET) imaging uses high-resolution radionuclides such as copper-64, bromine-76 and zirconium-89. Many tumour-associated antigens are now known and characterized, and antibodies to these allow targeting to different tumour types (Table 6.4).

Table 6.4 Some commonly used tumour-associated antigens with their representative antibodies

Antigens	Tumour type	Antibody
Tumour-associated glycoprotein 72 (TAG72)	Pancarcinoma	B72.3, CC49
Carcinoembryonic antigen (CEA)	Pancarcinoma	NP-4, A5B7
Polymorphic epithelial mucin (PEM)	Ovarian, breast, lung	HMFG1
Epithelial membrane antigen (EMA)	Colorectal and other epithelial tumours	17-1A
Epidermal growth factor receptor (EGFR)	Breast and lung cancer	425
P185 HER2/c-erb-B2	Breast and lung cancer	4D5
Prostate-specific membrane antigen (PSMA)	Prostate cancer	7E11-C5.3
CD33 67-kDa glycoprotein	Myeloid leukaemia	P67.6, M195
C20 35-kDa glycoprotein	Melanoma, neuroblastoma	C2B8
GD2 ganglioside		14–18

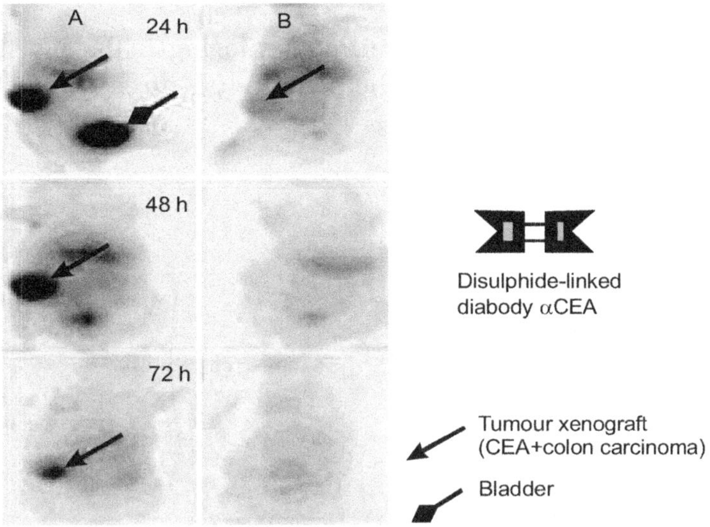

Figure 6.10 Imaging by αCEA bivalent diabody (A&B). At 24 h, radioactivity is seen in tumour xenograft and bladder and at 48 h and 72 h, it remains in colon alone. Control animals in the right are not showing radioactivity. (From Fitzgerald, *et. al.* (1997), *Protein Eng.* 10, 1221).

Antibody fragments, due to their smaller size, are capable of more readily penetrating solid tumours and should therefore show increased efficacy for imaging cancers. $F(ab)_2$, Fv and scFv show rapid clearance, and di-scFv and tri-scFv show improved results. Direct studies have confirmed their enhanced ability to penetrate tumours. Intact IgG molecules take 54 hours to travel 1 mm through a solid tumour, whereas Fab fragments can travel the same distance in 16 hours. The rapid clearance would be an advantage for *in vivo* diagnostic use, facilitating rapid removal of the conjugated radioactive tag but it is usually a disadvantage in a therapeutic application.

Bivalent diabodies are attractive molecules for tumour imaging (Figure 6.10) because of their increased valency and associated slow dissociation rates, and their rapid clearance from the circulation. Apart from specific and avid target binding, *in vivo* stability makes them more effective in imaging.

Strategies for Tumour Imaging

There are several strategies for tumour imaging. In the **one-step imaging** approach (Figure 6.11a), labelled antibody is directly injected into the patient.

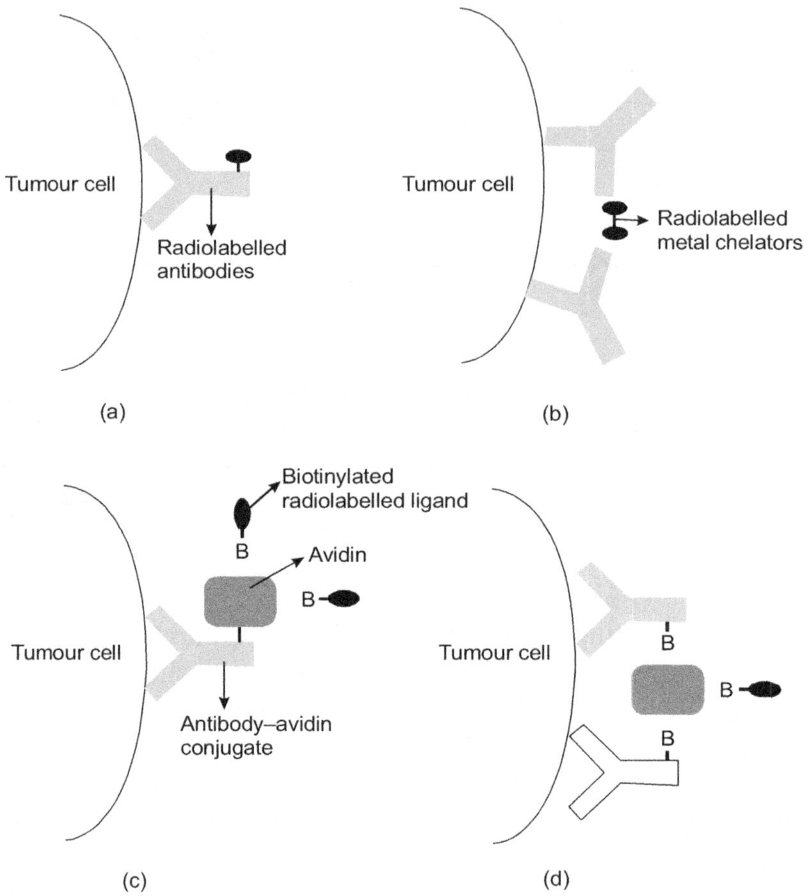

Figure 6.11 Different strategies for tumour imaging (a) One-step imaging (b) Two-step targeting (c) Strategy using antibody–avidin conjugate (d) Three-step strategy (King, D.J., 1998)

It circulates in the blood for a long period and builds up high levels of radioactivity at the tumour site. Good imaging can be achieved when enough radiolabelled antibodies reach the tumour site to generate a high "tumour : background" ratio. Imaging must be done days later when sufficient antibody has cleared from blood. Faster imaging is possible with antibody fragments. When the tumours are less accessible and less vascularized, early imaging with antibody fragments may be difficult and better results may be observed by imaging later with intact IgG.

In the **two-step targeting approach**, antibody is allowed to localize the tumour and sufficient time is allowed for antibody clearance from the blood and the non-target tissue. Radioisotope is injected separately in a form which can be readily captured by a tumour-bound antibody.

Bispecific antibody with specificity for tumour at one site and for a radiometal chelator like diethylene triamine pentaacetic acid (DTPA) at the other site, (Figure 6.11b) is injected first. After localization and clearance from blood, radiolabelled metal chelate is added. This is then bound by the antibody localized at the tumour site and rapidly cleared from the rest of the body. Divalent metal chelators improve imaging. Dia-Fab and tri-Fab constructs improve the image further. Bispecific antibodies have been successfully used to image thyroid carcinoma, colorectal tumours and non-small cell lung cancers.

Another strategy has also been developed. An antitumour antibody–avidin conjugate (Figure 6.11c) is injected, and is allowed to localize the tumour. Sufficient time is given to clear the unbound conjugates from blood. This is followed by the second injection of a low-molecular-weight biotinylated radiolabelled ligand that is captured by a tumour-bound antibody–avidin complex.

In the three-step strategy (Figure 6.11d), the biotinylated monoclonals are injected and allowed to bind to the tumour site. This is followed by the injection of avidin that binds with the tumour-attached biotinylated antibody. Radiolabelled biotin is then added, which is bound to the tumour site and is cleared from the rest of the body. This strategy improved sensitivity and also detected small tumour deposits that are not visible by other strategies.

IMMUNODETECTION OF OTHER DISEASES

Radioimmunodetection is not only used in tumour detection but also is used to detect myocardial necrosis, for imaging of blood clots and for detecting the infection and inflammation.

The ^{111}In-labelled Fab monoclonal fragment directed against heavy chain myosin is used to detect myocardial infarction, myocarditis, cardiac transplant rejection and cardiac involvement in systemic lupus erythematosus. Anti-fibrin antibodies are used for imaging blood clots both in deep vein thrombosis and arterial thrombosis. Antibodies specific for polymorphonuclear leucocyte surface antigen are developed for imaging inflammatory processes. ^{99m}Tc-Fab to the cell surface antigen NCA-90 has been used to detect soft-tissue infections and osteomyelitis. ^{99m}Tc-labelled anti-C15 antibody is used to image inflammatory disease.

Antibodies to CD4 are used to assess targeting in animal models of arthritis. Anti-CD4 reagents may also be useful in monitoring the distribution of CD4-positive lymphocytes in a HIV infection. Immunoimaging has also been useful in the detection of atherosclerotic plaques and lesions in Alzheimer's disease.

ABZYMES (IMMUNOCAPTURE)

It is possible to design an enzyme that detoxifies a drug in the bloodstream, destroys a virus, or targets the tumour cells, much the same way that a computer programmer designs a software programme. These custom-made enzymes are named as "abzymes" and are also called as "catmab" (from catalytic monoclonal antibody). Abzymes are usually artificial constructs, but are also found in normal humans (anti-vasoactive intestinal peptide autoantibodies) and in patients with the autoimmune disease systemic lupus erythematosus, where they can bind and hydrolyse the DNA.

The enzymes while functioning show a high-affinity interaction with the substrate, as a short-lived transition state. If antibodies are developed against the inhibitors that mimic this transition state, as their antigen-binding site, such antibodies can provide a catalytic function much more than the original enzyme. To say in simple terms, they are derived from a chemically stable hapten which mimics the translational state. Their binding energy enables the antibodies to increase both the specificity and the rate of a reaction. It has a tremendous potential to be used as a chemical tool.

In 1986, Peter Schultz and Richard Lerner demonstrated the generation of abzymes that catalyses ester hydrolysis—the breakage of an ester bond through the addition of water. The rates of reactions catalysed with abzymes, as measured by kinetic parameters such as K_M and V_{max}, are up to a millionfold greater than the corresponding uncatalysed reactions; however, in many cases, catalytic antibodies have not yet approached the rates of reactions catalysed by natural enzymes. To date there are more than 100 artificially generated monoclonal abzymes to catalyse more than 100 separate chemical reactions. They have been found to catalyse the hydrolysis of amides, esters, reactions for cyclization, decarboxylation, lactonization, photochemical thymine dimer cleavage, amide bond formation and other reactions that are not known to be catalysed by any known enzymes.

The following is a list of catalytic antibodies.

Esterase 2H6	Esterase 21H3
Oxidored.(hydride) 37B.39	Oxidored.(hydride) A5
Exo-Diels-Alderase 7D4	Endo- Diels-Alderase 22C8
Hydroxyl-THP synth. 26D9	Oxopane synth. 26D9
Thioester oxidored. 28B4	Esterase 17E11-115
Oxy-Cope isomerase AZ-28	Cyclopropanase 87D7
Enolase (multistriat.) 14D9	Cope-eliminase 21B12
Tandem cyclase 19A4	WMketone synthase 38C2
Naproxen esterase 5A9	Retroaldose (epothil.) 38C2
Xylulose synthase 38C2	Intra- and inter-amidase 14-10

Abzymes were initially applied to chemical processes which do not require catalysts. They are the antibodies with broad and programmable specificity. They are used to catalyse new chemical transformations. For example, various investigators have used catalytic antibodies to facilitate the Claisen rearrangement, drug degradation, prodrug activation, cationic cyclization, disfavoured ring closure, Diels-Alder reaction, aza-Diels-Alder reaction, oxy-cope reaction, double bond isomerization, asymmetric hydrogenation, oxidation, aldol reaction, retro aldol reaction and intramolecular Michael addition of aldehydes and ketones. One day the abzymes may find widespread applications in various fields.

The world's first commercially available catalytic antibody is the antibody 38C2. It has been shown to catalyse the aldol addition of a wide variety of aliphatic open chain and aliphatic cyclic ketones to various aromatic and aliphatic aldehydes. It is also involved in the condensation reaction, crossed aldol, retro aldol, self aldol, decarboxylation of β-keto acids, Robinson annulation and kinetic resolutions. More than 100 different substrate combinations—cross aldol and also intramolecular aldol reaction—have been identified. 38C2 is an efficient catalyst for retroaldol reaction, allowing the kinetic resolution of racemic secondary aldols. By using both the forward and backward, or synthetic aldol and retro aldol reaction, both the aldol enantiomers become accessible (Figure 6.12).

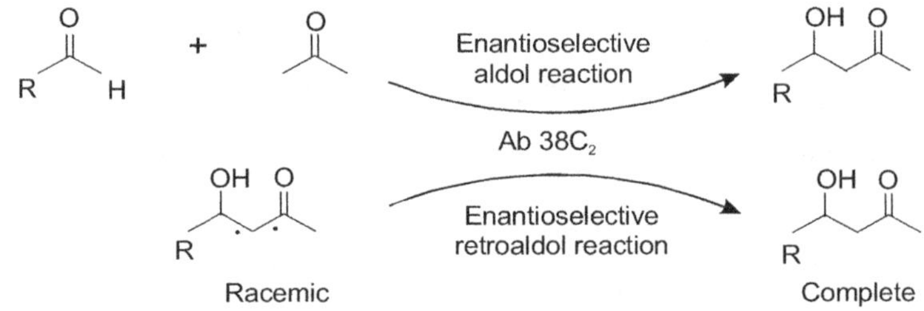

Figure 6.12 Role of antibody 38C2 in enantioselective aldol and enantioselective retroaldol reaction (www.scripps.edu/mb/barbas/antibody/sch1.jpg)

It is an excellent teaching tool in the laboratory for HPLC characterization of the antibody-catalysed reaction, for titration of antibody or enzyme-active sites, for analysing the kinetics of the antibody-catalysed reaction, Woodward UV rules, pK_a and enzyme mechanisms.

COCAINE DETOXIFICATION

Abzymes have been implicated for use in the detoxification of cocaine. Catalytic antibodies have been generated that cleave the cocaine molecule at specific bonds (Figure 6.13), thereby eliminating the toxic effect of the drug. As a pharmaceutical reagent, anti-cocaine abzymes could treat patients who are addicted to cocaine, or reverse the lethal effects of a cocaine overdose.

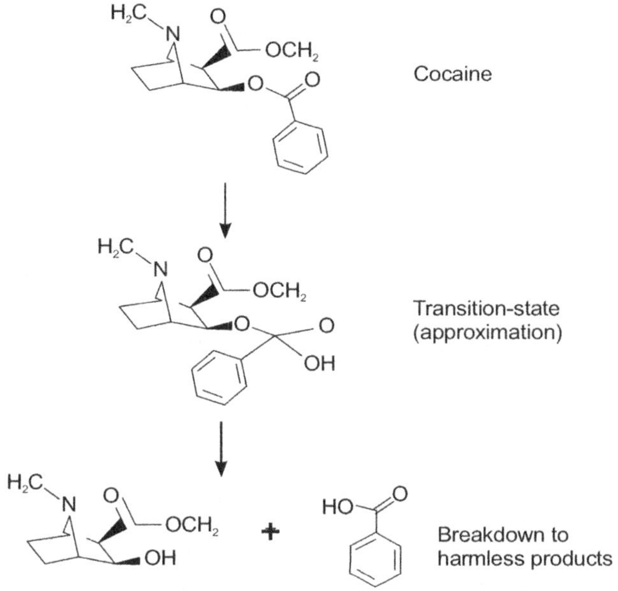

Figure 6.13 Role of anti-cocaine antibodies in the detoxification of cocaine (http://www.wiley.com/legacy/college/boyer/0470003790/cutting_edge/catalytic_ab/catalytic_ab.htm)

These abyzmes conferred the ability to degrade cocaine into the non-toxic ecgonine methyl ester and benzoic acid by-products before it reaches the central nervous system. Promising *in vivo* studies have demonstrated that the cocaine-degrading abzyme 15A10 is capable of protecting the rats that have overdosed on cocaine from seizures and sudden death, and cocaine reinforcing and toxic effects.

TREATMENT OF DISEASES

The function of an antibody is to protect the body from diseases. If the immune system fails to do that, then an antibody can be given as a therapeutic agent. The therapeutic application of an antibody has a long history. For many years, antisera from immunized animals or volunteer donors have been used to transfer immunity to patients who are vulnerable to infectious diseases. The application of new treatment modalities such as monoclonals is expected to enhance the therapeutic options available to treat infectious diseases. Monoclonal antibodies are offering substantial advantages in terms of potency, reproducibility and freedom from contaminants. They are hailed as **magic bullets** to specifically target cancer and other diseases.

New antibody pharmaceuticals offer novel and unique methods of treating some life-killing diseases and disorders. However, the very diversity and exquisite specificity of these new approaches pose huge new challenges to the pharmaceutical industry and drug regulatory mechanisms. Open-handed cooperation and collaboration in clinical research with experimental immunology during the past decades has brought the full benefit of antibodies in curing not only the genetic disorders like erythroblastosis foetalis (RhoGAM) but also to modulate and control autoimmune disorders. Table 6.5 shows the list of FDA-approved monoclonal antibody drugs.

The first monoclonals were murine antibodies, which when injected into a patient, the immune system identifies them as foreign proteins, and a human antimouse antibody (HAMA) response often occurs as mild allergic reactions in some patients or as severe reactions leading to death. In an effort to reduce this HAMA response, genetically engineered chimeric antibodies, which consist of the variable region of the mouse antibody combined with the constant region from a human antibody, have been developed. To reduce the HAMA response even further, humanized antibodies were developed. In these antibodies, only the complement-determining region (CDR) is retained from the murine antibody, and the rest of the antibody is human. More recently, some companies have developed methods to generate fully human monoclonals. Other approaches, including the use of antibody fragments, have also been used.

Three of the antibodies on the market in the United States that are approved for cancer, had more than $1 billion in sales each in 2005. Genentech developed all three of these blockbusters, and they are marketed by that company in the United States and by Roche in the rest of the world. They include Rituxan (rituximab, marketed as MabThera outside the United States) for treatment of non-Hodgkin's lymphoma, Herceptin (trastuzumab) for treatment of breast cancer, and Avastin (bevacizumab) for treatment of metastatic colorectal cancer. These three antibodies are included among Roche's five top-selling pharmaceutical products and had combined sales of greater than $6 billion worldwide in 2005.

Table 6.5 FDA-approved antibody drugs for various diseases

Year	Drug	Trade name	Target antigen	Treated for	Description	Product of
1986	Muronomab-CD3	Orthoclone-OKT3	CD-4	Transplant rejection	Murine Ig2a	Johnson & Johnson
1994	Abciximab	Reopro	Anti-GP IIb/III a	Cardiovascular disease	Chimeric Fab of IgG1	Centocor
1997	Rituximab	Rituxan	CD-20	Non-Hodgkin's lymphoma	Chimeric IgG1κ	Genentech
1997	Daclizumab	Zenapax	CD-25	Transplant rejection	Humanized IgG1κ	Roche pharmaceuticals
1998	Basiliximab	Simulect	α-chain of IL2 receptor	Transplant rejection	Chimeric IgG1κ	Novartis, Basel
1998	Trastuzumab	Herceptin	Her2 protein	Breast cancer	Humanized IgG1κ	Genentech
1998	Palivizumab	Synagis	RSV antigen	RSV infection	Humanized IgG1κ	Med Immune
1998	Infliximab	Remicade	TNF-α	Inflammatory diseases	Chimeric IgG1κ	Centocor
2000	Gentusumab ozagamicin	Mylotarg	CD-33 antigen	Acute myelogenous leukemia	Humanized IgG1κ conjugated immunotoxin	Wyeth–Madison
2000	Eculizumab	Soliris	C5 protein	Inflammatory diseases	Humanized antibody	Alexion pharmaceuticals

(Contd.)

Table 6.5 (Continued)

Year	Drug	Trade name	Target antigen	Treated for	Description	Product of
2001	Alemtuzumab	Campath-1H	CD-52 of B & T cell	Chronic lymphocytic leukaemia	Humanized antibody	Gen Enzyme
2002	Ibritumomab tiuxetan	Zevalin	CD-20	Non-Hodgkin's lymphoma	Murine radiolabelled	IDEC pharmaceuticals
2002	Adalimumab	Humira	TNF-α	Rheumatoid arthritis and psoriotic arthritis	Human antibody	Abbot
2003	Tositumomab	Bexxar	CD-20	Non-Hodgkin's lymphoma	Murine IgG2aκ radiolabelled	Corixa, Seattle
2003	Omalizumab	Xolair	IgE	Asthma	Humanized IgG1κ	Genentech
2003	Efalizumab	Raptiva	CD-11c	Psoriasis	Humanized IgG1κ	Genentech
2004	Cetuximab	Erbitux	EGFR protein	Colorectal cancer	Chimeric IgG1κ	ImClone Systems
2004	Bevacizumab	Avastin	VEGF	Colorectal cancer	Humanized IgG1κ	Genentech
2004	Natalizumab	Tysarbi	Anti-αk integrin	Multiple sclerosis	Humanized IgG4 κ	Biogen Indec
2006	Ranibizumab	Lucentis	VEGF	Inhibits angiogenesis	Humanized antibody fragment	Corixa, Seattle
2006	Panitumab	Vectibix	EGFR	Colorectal cancer	Human IgG2	Genentech, Novartis

Infectious Diseases

The monoclonal E5 raised against bacterial endotoxin (LPS) is used to control gram-negative bacteraemia. The first monoclonal successfully developed to combat an infectious disease is Palivizumab (Synagis) approved by FDA in 1998. It is used to prevent lower respiratory tract disease caused by respiratory syncytial virus (RSV). It is given as prophylaxis before infection. Another newly developed monoclonal cured the mice infected with West Nile virus. A single dose of antibody given soon after infection could boost the survival rates up to 90% or higher. Anti-HIV Fab fragments isolated by phage display library, block the infection by preventing the gp 120/CD4 binding event which is involved in the entry of the dreadful HIV virus. A number of sIgA that can be delivered orally, nasally or topically applied over vaginal mucosa for enteric, respiratory or STD respectively are under clinical trials.

While still in the early stages, other reports have indicated the possible uses of abzymes to inactivate viruses. For instance, abzymes that cleave the viral coat proteins of human immunodeficiency virus (HIV) have been isolated. Researchers have also developed abzymes that catalyse the specific destruction of viral genes. Perhaps in the future, we will have the tools to treat a wide variety of diseases through the use of catalytic antibody technology.

Cancer

Immunotherapy of cancer began about one hundred years ago, when Dr. William Coley at the Sloan-Kettering Institute, cured cancer using a mixed vaccine of streptococcal and staphylococcal bacteria, known as Coley's toxin. BCG vaccine developed in 1922 is even now used to treat bladder cancer, and as the stimulants of immune system. Many years of research have finally produced few "biological response modifiers" such as interferons, cytokines, vaccines and monoclonal antibodies for immunotherapy of cancer.

Monoclonal antibodies are used as target-seeking missiles or magic bullets that either can directly neutralize an offending agent or if equipped with a war head or poison arrow, can destroy the cancer cells. Antibodies raised against tumour-specific antigens are used for cancer treatment and also used to shrink tumours.

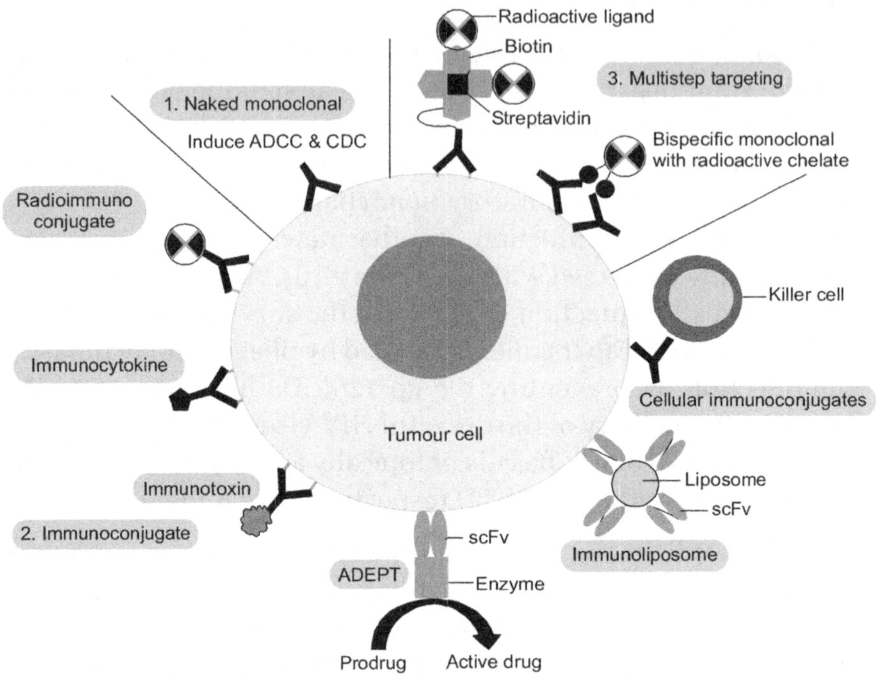

Figure 6.14 Schematic diagram showing antibody-based cancer therapy (*Nat. Rev. Cancer.* 2001. 1, 118–129.)

Monoclonals achieve their therapeutic effects through various mechanisms like ADCC (antibody-dependent cell-mediated cytotoxicity) and CDC (complement-dependent cytotoxicity). Various types of monoclonals used for such purposes are as follows:

- naked monoclonal antibodies
- conjugated monoclonals
- scFv
- bispecific monoclonals

Naked antibodies In tumour therapy, the unmodified or naked antibodies are used to stimulate the immune system through the effector function, such as ADCC or complement-mediated lysis (Figure 6.14). In some cases, a simple blocking or neutralizing effect is required. The other effector functions are given in Table 6.6.

Table 6.6 Effector function of antibody-targeted therapy

Function	**Mechanism**
Blocking/neutralizing	Antigen binding
Natural Fc-mediated effects	Complement fixation
	ADCC
	Phagocytosis
Cell signalling	Receptor cross-linking
Natural immune responses	Generation of anti-idiotype response
	Other 'vaccination' approaches
Artificial effectors	Radioisotopes
	Toxins—bacterial and plant
	Cytotoxic drugs
	Cytokines
	Enzymes—prodrug activation and direct toxicity
Bifunctional	Cross-linking cytotoxic effector cells
	Two-step targeting strategies for radioisotopes, toxins, etc.

The therapeutic monoclonal antibodies that are approved by the FDA are the following.

Campath (Alemtuzumab) A humanized antibody against CD52; destroys cancer cells by ADCC and CDC for advanced breast cancer.

Avastin (Bevacizumab) A humanized antibody against vascular endothelial growth factor (VEGF); prevents tumour angiogenesis of colorectal cancer.

Erbitux (Cetuximab) A chimeric IgG1 epidermal growth factor receptor; (EGFR) prevents ligand binding and affects cell signalling; used to treat colorectal cancer.

Vectibix (Panitumumab) A humanized monoclonal for epidermal growth factor receptor (EGFR).

Rituxan, Mabthera (Rituximab) A chimeric monoclonal against CD20; used to treat non-Hodgkin lymphoma.

Herceptin (Trastuzumab) A humanized IgG1 against ErbB2 (HER2); used to regulate over-expression of breast cancer.

Oncolym (Lym-1) Binds to the HLA-DR-encoded histocompatibility antigen that can be expressed at high levels on lymphoma cells.

Chemically cross-linked IgG constructs consisting of Fab fragments linked to two Fc regions to form Fab–Fc increase the therapeutic efficiency.

Conjugated antibodies These antibodies can be obtained by conjugating the monoclonals with any of the following:

- Radioactive atom
- Cytokine
- Plant and bacterial toxins
- Chemotherapeutic drug
- Abzyme (ADEPT)
- Liposome

The conjugated monoclonals are used as vehicles to deliver the drug or radioactive atom or toxin to the cancer cells. They act as a homing device, circulating in the body until it finds a cancer cell with the matching antigen. It delivers the drug or radioactive atom or toxin, near or inside the cancer cells minimizing the damage to normal cells. The FDA-approved conjugated monoclonals are the following.

Zevalin (Ibritumabtiuxetan) A monoclonal antibody against the CD20 molecule on B cells (and lymphomas) conjugated to either the radioactive isotope indium-111 (^{111}In) or the radioactive isotope yttrium-90 (^{90}Y) to treat the non-Hodgkin lymphoma patients.

Bexxar (Tositumomab) A monoclonal antibody against CD20 conjugated to the radioactive isotope iodine-131 (^{131}I). It is also designed to treat lymphoma patients.

CAT-3888 (formerly known as BL22) An antibody conjugated to pseudomonas exotoxin, a bacterial product that blocks the protein synthesis in the cells thus causing them to self-destruct by apoptosis. It is used to treat B-cell leukaemias and lymphomas.

Mylotarg (Gemtuzumab ozogamicin) A humanized monoclonal against CD33 an "immunoconjugate" linked to calicheamicin, a cytotoxic agent.

Radiolabelled antibody can kill cells from a distance and also kill cells adjacent to tumour, without the need for the penetration of the immunoconjugate into the tumour mass or without being internalized (Figure 6.15). Isotopes selected should have a high energy, low penetration and a short half-life and the products of radioactive decay should be inert. ^{90}Yttrium (beta-particle emitter) is the radioisotope which comes closest in fulfilling these needs. ^{111}Indium is the other one. α or β-emitting radionucleotides are used for the patients with hepatoma, human T cell leukaemia/lymphoma virus 1 and T cell leukaemia. Table 6.7 gives the list of radioactive materials used for conjugation.

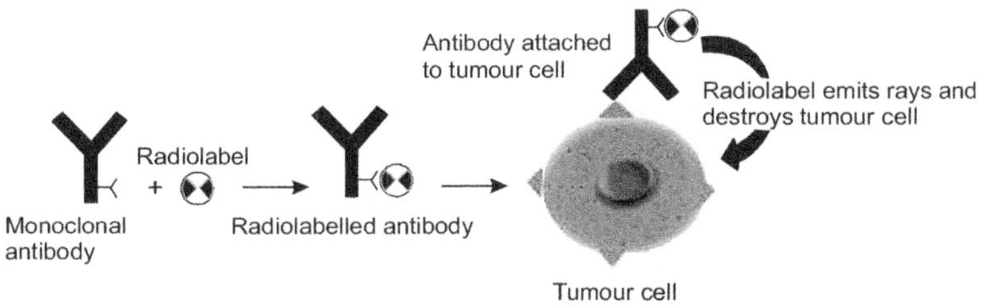

Figure 6.15 Radioisotope-material-labelled monoclonals in treating tumour cells

Table 6.7 Substances used to conjugate monoclonal antibodies

Chemotherapeutic drugs	Radioisotopes	Toxins
Methotrexate	^{131}I	Ricin
Daunorubicin	^{211}At	Abrin
Rubicin	^{90}Y	Pseudomonas toxin
Cisplantin	^{198}Re	Staphylococcal toxin
Vinca alkaloids	^{32}P	Diphtheria toxin
Adriamycin	^{212}Pb	Amantin, gelonin

A more promising approach is the use of certain powerful bacterial or plant toxin, where it is thought that a single molecule may kill a cell. Toxins of diphtheria, shigella and ricin (Figure 6.16a) consist of two polypeptide chains, joined by S–S bonds. Chain-B binds to cell surface, but chain-A is an enzymatically active species which enters the cell and destroys the protein-synthesizing machinery. B-chain in a toxin is replaced with specific monoclonal and targeted to specific cell for destruction (Figure 6.16b). This immunotoxin containing tumour-specific monoclonal antibody gets attached to a cancer cell, just like the B-chain of the toxin. It forms the endosome and releases the toxin into the cytoplasm. Diphtherial toxin inactivates the elongation factor 2 (EF-2) of the protein-synthesizing machinery. Hence, the tumour cell cannot synthesize proteins and is thus killed specifically.

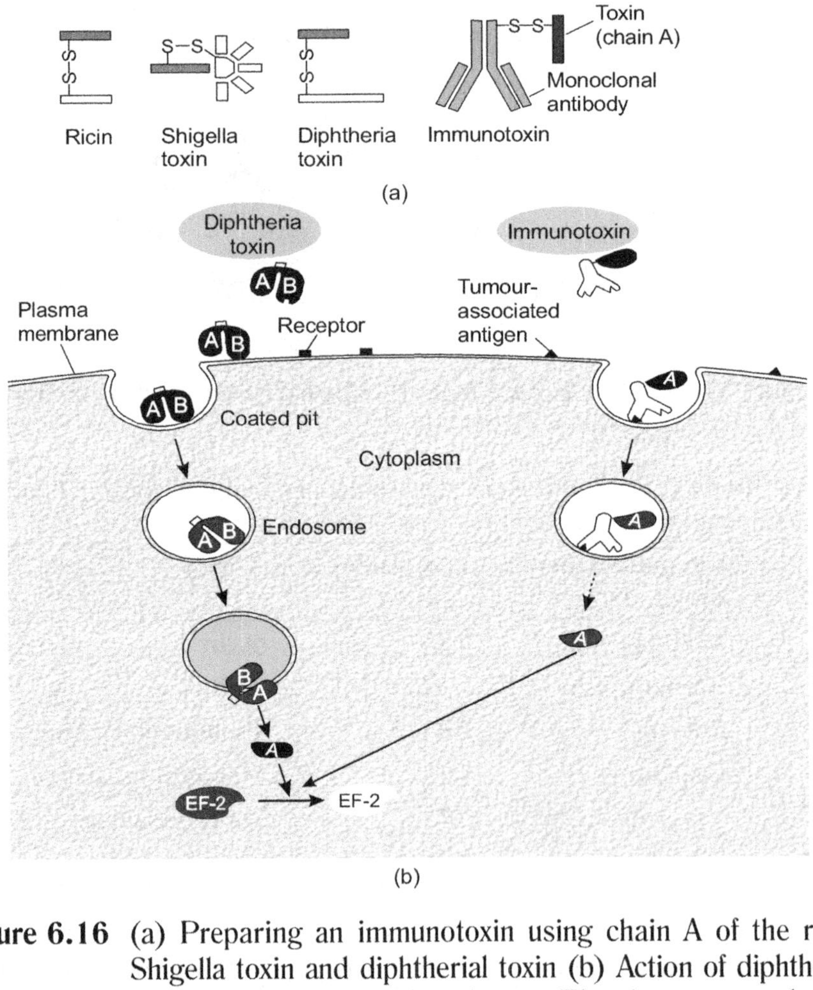

Figure 6.16 (a) Preparing an immunotoxin using chain A of the ricin, Shigella toxin and diphtherial toxin (b) Action of diphtherial immunotoxin and diphtherial toxin. They become attached to receptor, become internalized in an endosome, later get released and inhibit protein synthesis (Milstein, C., (1991)—Reading from *Scientific American*).

The use of conjugated monoclonals has several advantages. They will find their own way into the patients' body, will deliver drug to the target cells, and will attack only the target cells. The action is highly selective and so even a small dose of a drug or radioactive material is enough to kill the cancer cells. Apart from being used in treatment, radiolabelled antibody can also be used along with special camera to detect the areas of cancer spread in the body.

Bispecific antibodies Bispecific antibodies that can bind with their Fab regions both to the target (tumour) antigen and the effector cells, can be used for a safe, effective and rapid therapy. Designed to direct and enhance the body's immune response to specific tumours, these bispecific antibodies have shown promising results in cancer patients.

Diabodies The more potential, less immunogenic, short molecules called diabodies are used in the place of a whole antibody. Diabodies consist of a **targeting domain**, typically a fragment of a monoclonal antibody that binds to a cancer cell or an infected cell, linked to a **triggering arm** that is specific for a molecule capable of mediating a phagocytic or lytic response by the macrophages, natural killer cells, T-cells or other effector cells. Anti-HER2/neu diabodies are successfully treated for breast cancers. Bispecific diabodies with two different functional antibody-binding sites, αBCL1 and αCEA bind with the lymphoma target cell and CTL (Cytotoxic T lymphocyte)

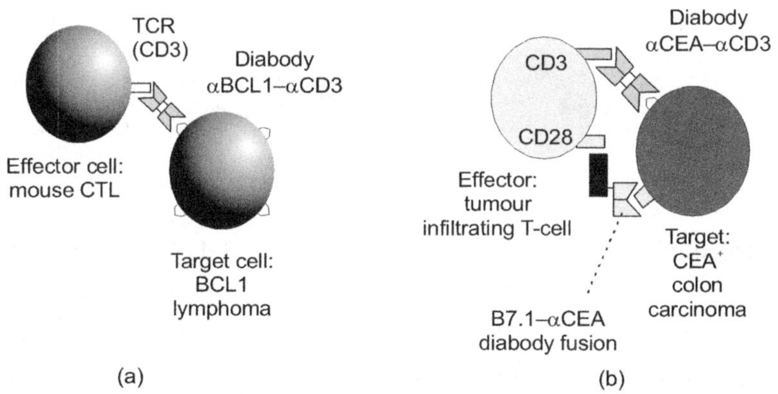

Figure 6.17 (a) Bispecific diabody for αBCL1–αCD3 (b) Two diabodies one for αCEA–αCD3 and another for αCEA-B7 (www2.mrclmb.cam. ac.uk/.../ images/Diab-24.2.gif)

simultaneously and thus allow a combination immunotherapy, focusing the effector arms of immune system on same target (Figure 6.17a). The two signals can be delivered by two diabody constructs, a bispecific diabody engaging CD3 and the target cell as well as a B7.1–αCEA diabody fusion protein that stimulates CD28 of a T cell through co-stimulatory signal (Figure 6.17b).

Diabodies engineered to cross-link the target cells and serum immunoglobulin (IgG) become "antibodies by proxy", acquiring a whole spectrum of antibody effector functions such as complement recruitment, ADCC, phagocytosis, superoxide burst, T-cell activation, prolonged serum half-life and antigen presentation (Figure 6.18a). Ig diabodies in conjunction with T-cell recruiting diabodies may thus allow a **combination immunotherapy** focusing both the effector arms of the immune system on the same target cell (Figure 6.18b). The di-diabody mediates effective antigen-dependent cell cytotoxicity, and was as effective as the combination of the two monospecific parent antibodies. It should be effective in patients with a wide range of cancers.

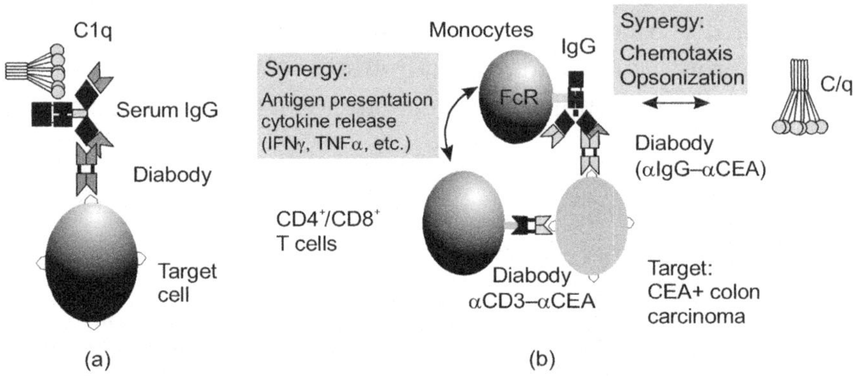

Figure 6.18 (a) A diabody attached to the target cell triggering the immune arm of the body (b) Synergic action of IgG diabodies and T-cell recruiting diabodies—combination therapy

By mediating an immune assault on pathogens, diabodies may also lead to the antigen presentation and a vaccine-like response in patients. Over the next few years, we expect several bispecific diabodies and triabodies (Figure 6.19a and b) to enter the late stages of clinical trials.

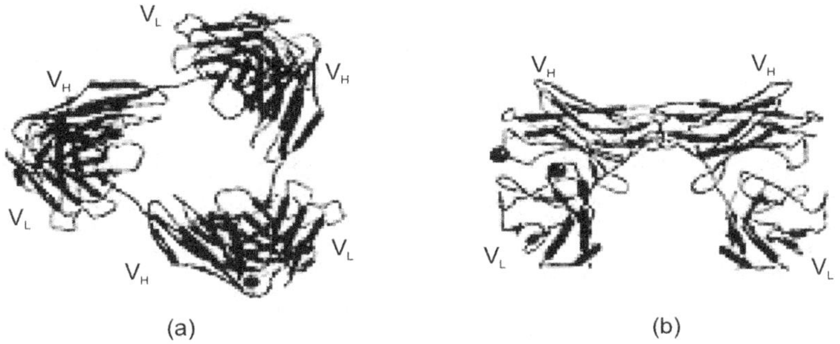

Figure 6.19 (a) A triabody with three V_H/V_L regions (b) Ribbon structure of diabody

ADEPT ADEPT is the "antibody-directed enzyme prodrug therapy". It is a new type of site-specific targeted therapy for cancer, involving both an antibody–enzyme conjugate and a low-toxicity prodrug. It is the most exciting application of abzyme technology. By utilizing antibodies that specifically bind to tumour cell antigens, cancer drugs can be delivered directly to the tumour. One site of abzyme binds with high affinity to a tumour cell antigen, while the second site catalyses the cleavage of a prodrug. First, the monoclonal–enzyme conjugate is administered into the patients. It binds the tumour cells with high affinity. The unbound antibody is allowed to clear from the body. Second, the prodrug is introduced into the bloodstream. The prodrug comes into contact with the enzyme; a reaction takes place, which activates the anticancer drug by cleaving. By this technique, tumours are selectively destroyed while healthy cells are spared from the toxic effect of cancer drugs (Figure 6.20). Not all tumour cells need to be targeted, because activated drug accumulates and diffuses at the tumour site, killing the nearby cells.

Abzyme 38C2 is used to unmask the prodrugs of doxorubicin and camptothecin when they are applied at therapeutically relevant concentrations. These prodrugs have substantially reduced toxicity due to their masked active groups. 28B4-abzyme catalyses the periodate oxidation of *p*-nitrotoluene methyl sulphide to sulphoxide.

The problem of ADEPT is to identify an enzyme that is not already present in human. To overcome this pitfall, non-human enzymes, such as those found in bacteria, have been used. However, the immunogenicity associated with using these foreign proteins limits their clinical value. In their studies, Blackburn and his colleagues devised a bipartate carbamate prodrug that could be acted upon by antibodies in a manner similar to the bacterial enzyme CPG2 exopeptidase, previously shown to convert the prodrug into its cytotoxic form. They generated several abzymes capable of hydrolysing the carbamate ester bond in the nitrogen mustard prodrug into a compound with cytotoxic activity.

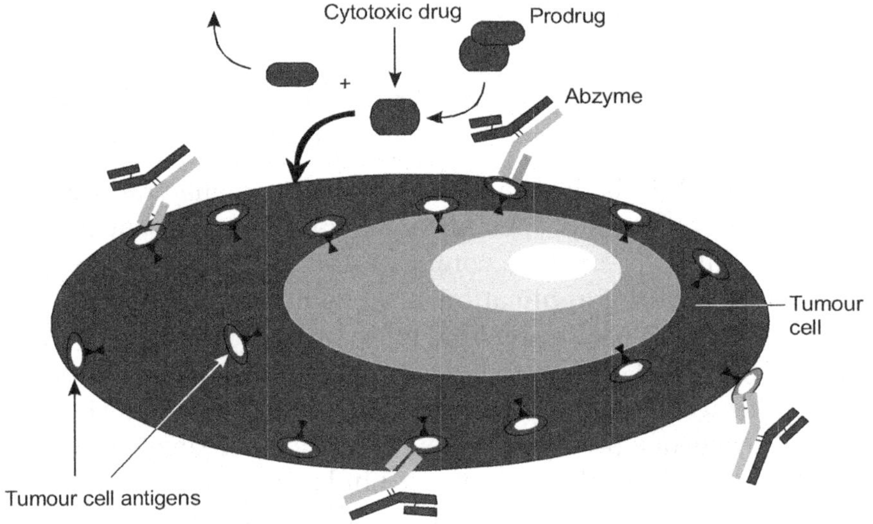

Figure 6.20 Bispecific abzyme 38C2 unmasks the cancer prodrug by its catalytic site while it is attached to tumour-specific anigen by its other Fab end (www.utc.fr/~friboule/ images/figures/ miniadapt.gif)

Bone marrow transplants Monoclonals are also used to cleanse the bone marrow, the source of many blood-related cancers. Some bone marrow is removed and infused with millions of magnetic beads, each coated with monoclonals specific to the cancer cells. When the marrow is passed through a magnet, the cancer cells are pulled out. The clean marrow is then returned to the patients.

Autoimmune Diseases

Monoclonals are used to predict or treat autoimmune diseases. HLA-typing identifies the persons at risk of particular autoimmune diseases and helps them to take preventive measures. Persons with HLA-D3 antigen are 73 times more prone to colic disease and with

HLA-D2 antigen is 4 times more prone to rheumatoid arthritis. HLA-B27 antigen is found in 80–90% of people with ankylosing spondylitis and Reiter's syndrome, and can aid in the diagnosis of these diseases. Since HLA-B27 is also present in 5–7% of people without autoimmune disease, the mere presence of this HLA molecule is not indicative of the disease.

A second major therapeutic area in which monoclonals have demonstrated both the clinical and commercial success is the immune-related diseases. These diseases include autoimmune diseases, inflammation, and immune suppression (for preventing the rejection following transplants). The first FDA-approved monoclonal drug for inflammatory diseases is infliximab (anti-TNFα antibody) used to treat rheumatoid arthritis. Eculizumab was approved in 2000. It is a humanized monoclonal that prevents the cleavage of human complement component C5 into its pro-inflammatory component. It is named as Alexion and is used for treating inflammatory diseases like rheumatoid arthritis and nephritis.

Two "blockbuster" monoclonals, with annual sales that exceed $1 billion each, targeting the immune system are Centocor's Remicade (infliximab) and Abbott's Humira (adalimumab). Both of these drugs target the tumour necrosis factor alpha (TNF-α), although Remicade is a chimeric antibody and Humira is a human antibody. Remicade is indicated for treatment of rheumatoid arthritis, Crohn's disease, ankylosing spondylitis, psoriatic arthritis, and ulcerative colitis. Humira is indicated for treatment of rheumatoid arthritis and psoriatic arthritis. Both of these antibodies compete against Amgen's Enbrel (etanercept), made in the TAC, which is not a true monoclonal. It is a fusion protein that consists of the extracellular ligand-binding portion of tumour necrosis factor receptor (TNFR) linked to the Fc portion of human IgG1. The chimeric construct had a very high affinity for TNF and was able to block this critical cytokine *in vivo*. The role of anti-TNF agents is now widely recognized and several are available on the market. Enbrel, a monoclonal against tumour necrosis factor (TNF), is used for the treatment of rheumatoid arthritis, polyarticular-course juvenile rheumatoid arthritis, ankylosing spondylitis, psoriatic arthritis and psoriasis.

Clinical trials in autoimmune diseases were pioneered at Addenbrooke's hospital, Cambridge, in the departments of

Rheumatology, Medicine and Neurology. Several antibodies have been tested, including CAMPATH-1H (for depleting lymphocytes), antibodies against CD25 (specific for activated T cells), CD4 (for blocking the function of the critical helper T cells) and CD18 (which blocks the migration of leucocytes from the blood to sites of inflammation). Anti-CD18 antibody reduces inflammation and tissue injury in rheumatoid arthritis.

Efalizumab blocks one of the receptors called leucocyte function antigen-1 or LFA-1 present on the T lymphocyte. By blocking the adhesion, efalizumab selectively and reversibly blocks the activation, reactivation and trafficking of T cells that lead to the development of psoriasis.

Allergy

Allergy is due to high IgE level. Allergy can be treated with anti-IgE antibodies. Omalizumab (anti-IgE) is a very high-affinity recombinant humanized monoclonal that not only reduces the symptoms of asthma, hay fever and other allergies, but also decreases the use of corticosteroids.

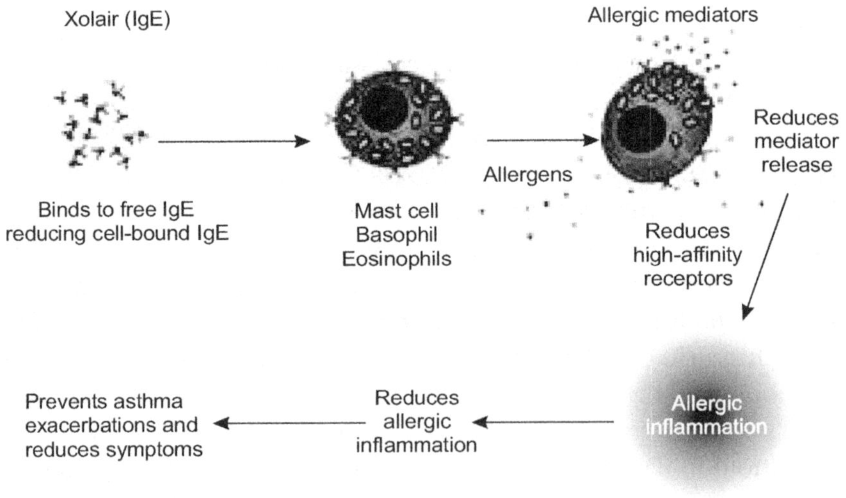

Figure 6.21 Mechanism of Xolair in preventing persistent asthma (www.fda.gov/.../ image002.jpg)

Omalizumab is given as subcutaneous injection. It is approved by FDA in 2003. It is also named as Xolair. The mechanism of action

(Figure 6.21) differs completely from standard treatments for persistent asthma. It is designed to bind to the circulating IgE antibodies in the blood, decreasing the amount of IgE antibodies available to bind the mast cells. With Xolair, fewer IgE antibodies can bind to mast cells, making IgE cross-linking less likely and inhibiting the mast cell's release of those chemicals that can cause inflammatory responses in the body.

Cerebral Thrombus

Monoclonals prevent the arteries from reclogging in people with a history of cerebral thrombus. Plasmin and fibrin are responsible for the degradation of clot (Figure 6.22a). Anti-fibrin monoclonal chemically coupled with tissue-specific plasminogen activator, when injected, bound to blood clots. After binding to fibrin of blood clot, the plasminogen activator causes the plasmin to accumulate in the vicinity of the clot. The plasmin then degrades the clot (Figure 6.22b). It dissolves the clot without any significant fibrinogen breakdown.

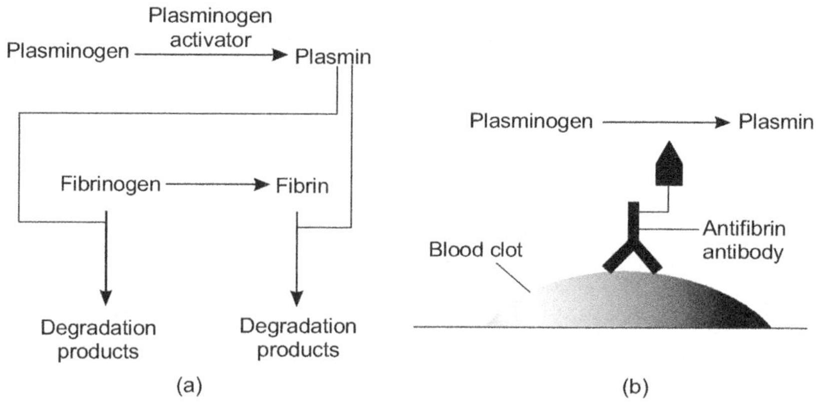

Figure 6.22 (a) Role of plasmin and fibrin in the degradation of clot (b) Role of anti-fibrin monoclonal coupled with plasminogen activator in removing the clot

Cardiovascular Disease

A heart attack may occur when a blood vessel in the heart is blocked by a blood clot. Abciximab (ReoPro) is a chimeric monoclonal, used to lessen the chance of heart attack in people who need percutaneous coronary intervention, a procedure to open the blocked arteries of the heart. This monoclonal is directed against platelet glycoprotein IIb/IIIa receptor.

Diabetes

An aglycosyl CD3 antibody has been currently tried for the treatment of type I diabetes. By treating the patients very early after diagnosis, it may be possible to arrest the process of destruction of insulin-producing cells and thereby to enhance the quality of their life subsequently.

The Therapeutic Antibody Centre (TAC) was set up by Geoff Hale and Herman Waldmann to provide a way of making a range of therapeutic monoclonal antibodies available for clinical trials in order to explore the basic principles of antibody therapy and provide the groundwork for later commercialization of useful products. Overall, more than 1 kg of antibodies has been made and over 5000 patients treated, which is more than many substantial biotech companies can boast of. Much of the early work was with CAMPATH-1H antibodies, widely used for prevention of graft-versus-host disease and graft rejection in patients receiving bone marrow transplants and now licensed for treatment of leukaemia.

Another important principle of the TAC is the ability to take the problems identified in the clinic back to the research lab in order to provoke a fresh cycle of research and development. This iterative process is now leading to a "second generation" of therapies which we believe will give us short-term treatments with a long-term impact.

ANTI-IDIOTYPIC VACCINES

When an antigen with epitope 'E' is injected in animal 'A', anti-E antibody can be obtained. When this anti-E-antibody is immunized in animal-B, it produces anti-anti-E-antibody which is also called as anti-idiotypic antibody (Figure 6.23). This antibody mimics the epitope E. Anti-idiotypic monoclonals have been evaluated as (they mimic) immunogens for their active immunization and are used as vaccines successfully for human sleeping sickness, *Trypanosoma* infection in mice and *Eimeria* infection in chickens. Table 6.8 shows the list of anti-idiotypic vaccines for infectious diseases and cancer.

Anti-idiotypic monoclonal antibodies can be used as part of a cancer specific vaccine because they look like the antigens that are originally found on the cancer cells. Therefore, they trigger an immune response against specific cancer. Like antigen vaccines, anti-idiotypic vaccines are not unique for each patient. Anti-idiotypic antibodies are tested for melanoma, prostate, breast, pancreatic and lung cancers, soft tissue melanoma and other cancers. Preliminary studies of B cell lymphoma vaccine have yielded promising results.

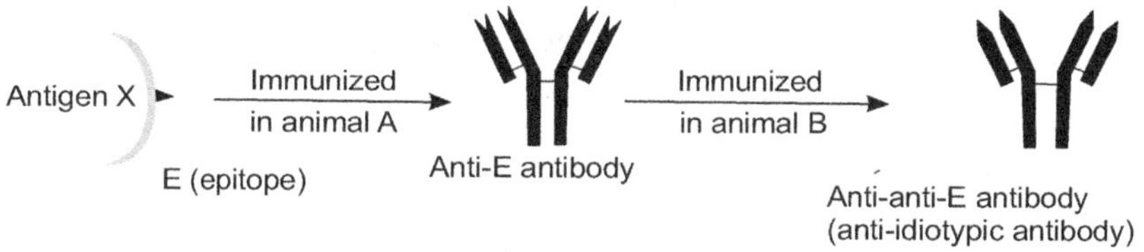

Figure 6.23 Production of anti-idiotypic antibody

Table 6.8 Diseases for which anti-idiotypic vaccines are available

Viral infections	Bacterial infections	Parasitic infections	Cancer
Polio	*Listeria monocytogenes*	*Schistosoma mansoni*	17-A
Reovirus		*Trypanosoma cruzi*	105 AD7
Rabies	*E. coli*	*Trypanosoma rhodesiense*	CEA vaccine
Hepatitis	*Streptococcus pneumoniae*		IE-10
Sendai virus			

TRANSPLANTATION

Tissue Typing

HLA (human leucocyte antigens) are proteins found in the membranes of nearly every nucleated cell in the body. These antigens are in especially high concentrations on the surface of leucocytes. HLA antigens are the major determinants used by the body's immune system for recognition and differentiation of self from non-self. There are many different major histocompatibility (MHC) genes coding HLA proteins, and individuals possess only a small, relatively unique set that is inherited from their parents. It is unlikely that 2 unrelated people will have the same HLA make-up. Hence tissue typing is essential prior to organ transplantation in order to avoid adverse rejection episodes.

Tissue typing involves the identification of MHC class I antigens on both the donor and recipient cells (WBC). Monoclonals are used in tissue typing. By matching MHC alleles of the donor cells identical to those of recipient cells, the closest match of the donor organ is selected for transplantation. Matching MHC Class II (HLA-DR) allele is more important than Class I. Monoclonals against different HLA-antigens (at least six antigens) permit better tissue or lymphocyte typing of donor and recipient to enable better cross matching.

Immunosuppressive Drug

By injecting a specific antibody prior to transplantation of kidney or bone marrow, one can inhibit allograft rejection, inflammatory and cytotoxic responses. Graft rejection is caused principally by a cell-mediated response to alloantigen. From the donor kidney, passenger leucocytes (APC) migrate to regional lymph nodes, resulting in the activation and proliferation of T helper cells, which via IL-2, activate T-delayed hypersensitive cells and cytotoxic T cells. These cells mediate graft rejection (Figure 6.24). To reduce the episodes of kidney rejection, anti-CD3 (Orthoclone OKT3 or Muromonab-CD3) antibodies (raised against CD3 of T helper cells) are used. It binds with CD3 and inactivates functions of T cell and thus results in immunosuppression. Orthoclone OKT3 reverses graft rejection, probably by blocking the function of T cells and effector cells.

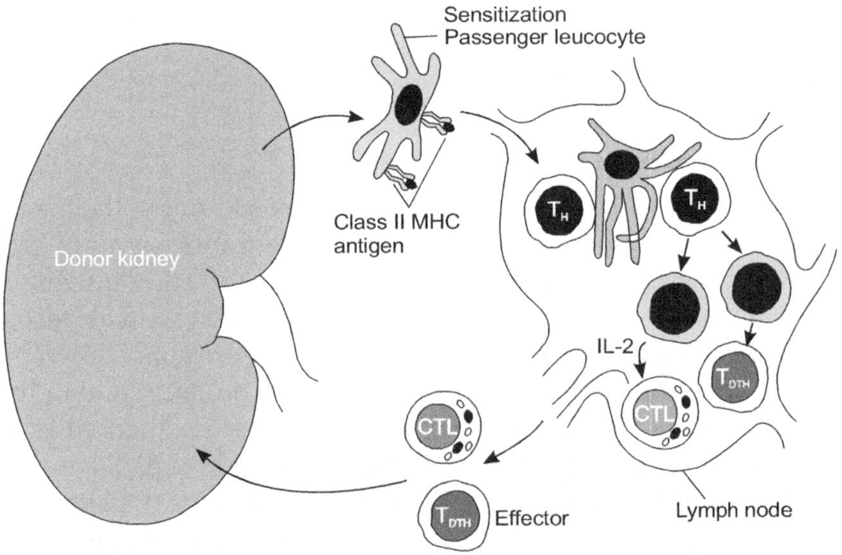

Figure 6.24 Diagram showing allograft rejection (Kuby, J., 1997)

Orthoclone OKT3 is a murine monoclonal, first FDA-approved immunosuppressive pharmaceutical with less side effects than other immunosuppressive agents for treating acute organ transplant rejection. 8 years elapsed before the next therapeutic monoclonal was approved by the US Food and Drug Administration. One factor that contributed to this gap was the observed immunogenicity of mouse antibodies in human patients, which can lead to rapid clearance, reduced efficacy and an increased risk of infusion reactions, which can range from relatively benign fevers and rashes to cardiopulmonary and anaphylactic-like adverse events.

Following this, Basiliximab Simulect and Daclizumab (Zenapax) were developed. They are more humanized monoclonals developed against IL-2 receptor and approved for the same treatment.

ENVIRONMENTAL PROTECTION

Monoclonal antibodies are indirectly helpful in environmental protection. They are used to identify toxic pollutants and harmful biological reagents in environmental and food samples.

Dioxins are organic compounds that have two benzene rings linked which include PCDD (polychlorinated dibenzodioxin), PCDF (polychlorinated dibenzofuran) and PCB (polychlorinated biphenols). There are 17 toxic dioxins whose concentrations in parts per trillion levels are measured by several assays. The time-consuming (several weeks), costly ($1000 per test) tests are replaced by simple, cheap and faster EIA methods that use specific monoclonals linked with horseradish peroxidase enzyme (Figure 6.25). This assay was developed in 1987. The dioxins are detected even in complex samples like beef fat, chicken fat, pork fat, eggs, milk powder, composite animal feeds, and human serum. A monoclonal antibody against surface antigen of *Sphingomonas* sp. strain RW1 is used to detect the organism in environmental samples by epifluorescence microscopy, which is further confirmed by western blot. This organism is capable of degrading dibenzo-*p*-dioxin.

The environmental protective agency in 1996 gave guidelines for the need for appropriate species-specific monoclonal in cell phenotypic assay. Monoclonals have been used to detect potential carcinogens in nanogram amount in environmental samples. A rapid and accurate *in vitro* immunofluorescent test uses high-affinity mouse monoclonals for simultaneous detection of faecal contaminations of *Cryptosporidium* oocysts and *Giardia* cysts in water and environmental samples.

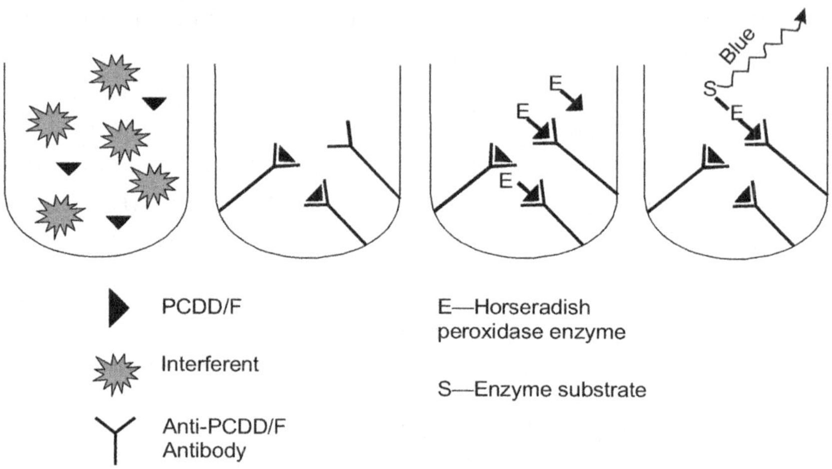

Figure 6.25 Enzyme immunoassay for detecting dioxin (Marcus, *et al.*, 2000)

Scientists in the Government laboratories have produced antibody-based biosensors that detect explosives at old munition sites. These methods are cheaper and faster than current laboratory methods that require large and expensive instruments. As these tests include portable instruments, rather than gathering soil sample and sending them to laboratories for analysis, scientists can measure the level of contaminants on the site itself and know the results immediately.

IMMUNO(BIO)SENSORS

The immunosensors are the alternatives to immunoassays for monitoring antigen–antibody reactions in real time, without the use of labelling. An immunosensor is a solid-state device in which the antigen–antibody reaction is detected via a transducer which provides a signal during the reaction. Three main types of transducers have been used in the biosensor technology. These exploit changes in mass (piezoelectric or acoustic wave), electrochemical properties (potentiometric, conductometric or amperometric) or optical properties (fluorescent, luminescent, surface plasmon resonance or waveguide properties).

Mass Detecting Immunosensors

In the piezoelectric biosensors, quartz crystals are coated with antibody. Binding the antigen results in a change in the mass which is monitored through a change in the frequency of the crystal oscillation. With the use of such sensitive devices, measurement of insulin, metamphetamine detection in human urine, herbicide detection in drinking water and detection of viruses and bacteria are possible.

A further variant of the piezoelectric sensor uses surface acoustic waves to monitor the binding of sample antigen to the antibody attached to quartz crystal. In this application, the oscillation of the crystal is at higher frequency and an acoustic wave is generated by application of an alternating voltage across interlaced electrodes, known as an interdigital transducer. A second interdigital transducer detects the acoustic signal a few millimetres away. Binding of sample to the crystal slows the acoustic wave, with a change in the velocity being proportional to the analyte concentration. Such devices have the potential for higher sensitivity but may suffer more interference from factors such as temperature, pressure and conductivity, all of which may affect the properties of the acoustic wave.

Electrochemical Immunosensors

Potentiometric immunosensors are based on the change in potential that results when an antibody is immobilized on an electrode and its antigen binds to it. Antibodies like all proteins are polyelectrolytic and in many cases binding of antigen will alter the charge. Therefore, the potential difference between an electrode with immobilized antibody and a reference electrode will depend on the concentration of antigen present. However, direct detection results in only small changes in the potential.

Conductometric sensors monitor variations in the conductivity between two electrodes by measuring variations in the current across the electrodes. To date, application of this technology to immunosensors has been limited. Amperometric sensors measure the current generated when electroactive species are reduced or oxidized at the electrode. Antibodies are not directly electroactive and thus enzyme labels are used to generate electroactive species. Such sensors often use oxygen or hydrogen peroxide electrodes with the current produced being directly proportional to the amount of oxygen or hydrogen peroxide reduced or oxidized.

Optical Immunosensors

Optical immunosensors can be used with or without enzyme labels. Antibody-labelled immunosensor based on fluorescence or luminescence is attractive due to the relatively simple instrumental design. The most widely used optical immunosensors do not require labelled reagents.

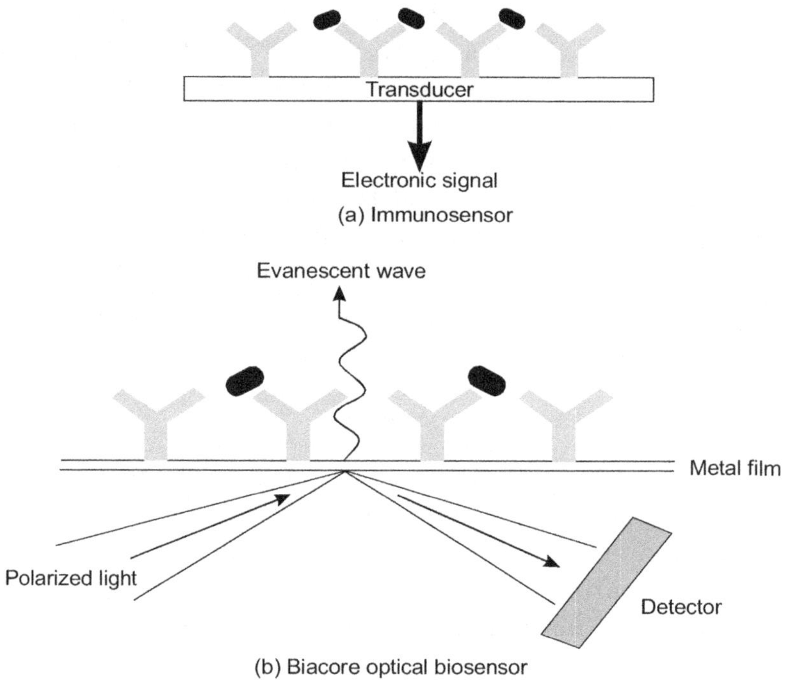

Figure 6.26 Diagram showing the optical immunosensors (King, D.J., 1998)

In the surface plasmon resonance (SPR), a metal film coated with a dextran layer is used to immobilize the antibody or antigen. When light hits the metal film, surface plasmons in the film are excited and an evanescent wave is generated, a process known as SPR, which results in a decrease in the intensity of reflected light. When an antigen binds to the immobilized antibody, the resulting change in the refractive index causes a change in the angle at which the drop in reflected light occurs.

Continuous monitoring of the angle and intensity of reflected light therefore allows the real-time analysis of binding of events. A commercial system based on this principle, known as the Biacore, is now widely used. In this system, SPR technology has been combined with a microfluidic system such that continuous monitoring is possible (Figure 6.26).

Another commercially available type of optical biosensor is based on a combination of SPR technology with wavelength technology. This device has found application in the area of bioprocess monitoring. The production of Fv fragment could be monitored through fermentation and purification processes. Measurements were obtained in five seconds, with an overall assay time of two minutes.

Two insecticidal proteins, *Bacillus thuringiensis* delta endotoxin (Bt toxin) and proteinase inhibitor I (PI-I), and a bactericidal polypeptide, cecropin, have been used as model systems to investigate the feasibility of monitoring transgenic plant pesticides in the environment. Since, after harvest, much of the plant residue that contains the transgenic product may be incorporated into the soil, the fate of the pesticidal compounds in soil is of particular interest. Methods were developed to extract the pesticidal proteins from soil. ELISA or bioassays were adapted or developed to quantify the pesticidal proteins recovered from soil. Detection limits are in the 1–10 ppb range.

DETECTION OF TOXINS, RESIDUES AND CONTAMINATIONS IN FOOD AND FEED

Monoclonals are used to detect toxins in food and animal feed. "Toxiklon" is a commercial mycotoxin kit used to test the animal feed by ELISA. Fusarium toxins like T2, Zearalenone, DON Fumonisins and ochratoxin-A and also aflatoxin-A are detected by monoclonal-based ELISA. Monoclonals are avialable to diagnose bacterial toxins such as *Bacillus cereus* diarrhoeal toxin, *Clostridium botulinum* toxin, C. *perfringes* enterotoxin, Shiga toxin and contaminations of a number of bacteria like *Campylobacter, E. coli, Listeria, Salmonella, Staphylococcus aureus* and *Vibrio cholerae* in the food. To detect the presence of *Salmonella enterica* and *Helicobacter pylori* in food and environmental samples, magnetic capture beads are employed.

Monoclonals are also used to detect residual drugs in animal food products like milk, meat, honey, egg or sera. In the farm animals, antibiotics are added in their feed, as a feed additive, to increase their weight and also as a prophylactic treatment to avoid sickness. These additives leave toxic residues in animal tissues and their products. These residues pose health risk and undesirable effects like allergy, toxicity, mutagenicity and development of antibiotic resistance in human and other animals. The maximum drug residue limit (MRL) in these products is detected using monoclonals against chloramphenicol, gentamcin-BSA, sulphanamide and other antibiotics.

RESEARCH

Monoclonals are extremely valuable, powerful, versatile and indispensable tools both in research and in discoveries. They are used to identify the enzymes, cell secretion, change in surface mapping, surface receptors, molecular events involved in signal transduction, the study of cytoskeleton, etc. Without the monoclonals, numerous discoveries would not have been possible.

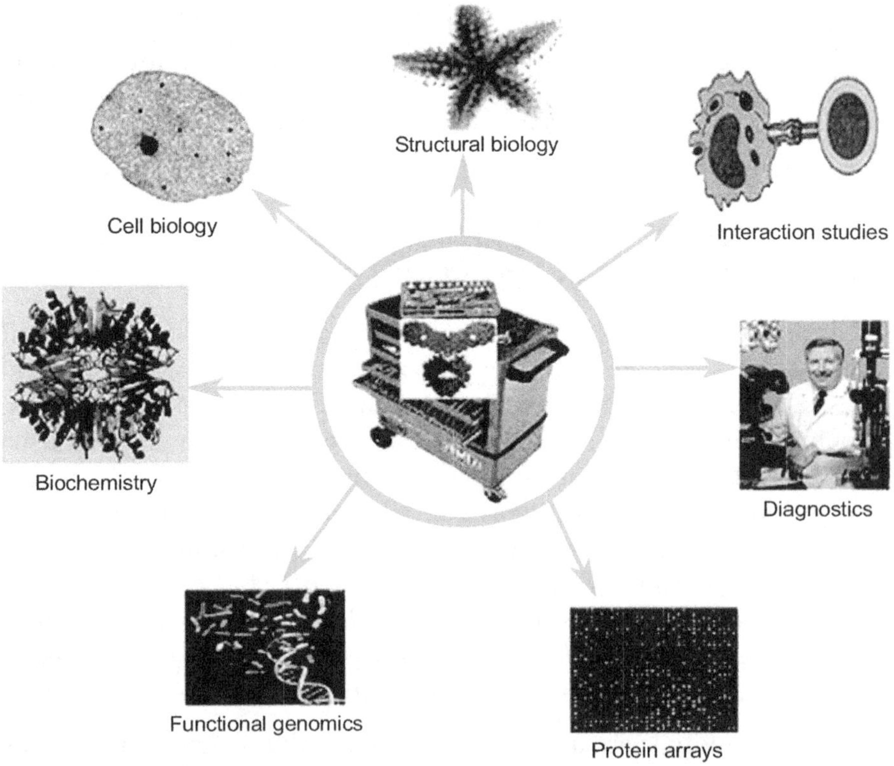

Figure 6.27 Diagram illustrating the use of monoclonals in research

One of the most exciting investigative applications of monoclonals is in the area of membrane research. Membrane proteins are hard to purify. They are present in cells in small amounts, often they have no easily measured biological activity or else their activity is destroyed when the membranes are solubilized for analysis. One way to overcome these problems is to characterize cell-surface molecules by antibody technique.

In 1977, Galfre and Milstein along with Alan F. Williams of University of Oxford immunized a mouse with cell membranes from rat thymus. After immortalizing the mouse cells by fusion, they were able to isolate the clones producing different antibodies. Since then, a number of other monoclonals to differentiate antigens of mouse, rat and human cells have been prepared. This approach led to the discovery of cluster differentiation (CD) for human lymphocyte, subsets of T cells, interaction of T cells with APC, molecular events involved in the processes, intracellular components of signal transduction and other biochemical pathways that are active during immune response. Antibody coupled with fluorescent dyes has helped a lot in studying the cytoskeleton under UV glow.

Monoclonals are essential reagents for isolation, identification and cellular localization of specific gene products and the transcription factors, and for assisting in the determination of their macromolecular structures. It is the facility to make monoclonals against the recombinant proteins or peptides based on the protein-sequence-derived cDNA clone that has made an outstanding contribution to our understanding of cell biology. Thus it is an indispensable tool for proteomic and genomic research (Figure 6.27).

Determination of Protein Structure

Monoclonal fragments help in determining the protein structure by X-ray crystallography. Many proteins of biological structure are unable to form the high-quality crystals required for structural determination due to heterogeneity, insolubility, flexibility or polydispersity in the solution. The soluble Fab and Fv fragments bind tightly to these protein antigens, effectively transform aggregated proteins into soluble monodisperse samples suitable for crystallization. The HIV capsid protein (p24) was able to crystallize only as a Fab–p24 complex. HIV reverse transcriptase allows rapid determination of structure when used along with an Fab fragment that decreases the mobility. Bacterial cytochrome C oxidase, a membrane protein, shows enlargement of the polar surface to form crystal lattice when it combines with the Fv fragment. Selection of appropriate antibody fragments from phage display may allow rapid identification of suitable reagents.

Purification of Proteins

Before the advent of monoclonals, the purification of minor protein components from a complex mixture often required numerous chromatographic steps and generally had low yields. Once the monoclonal

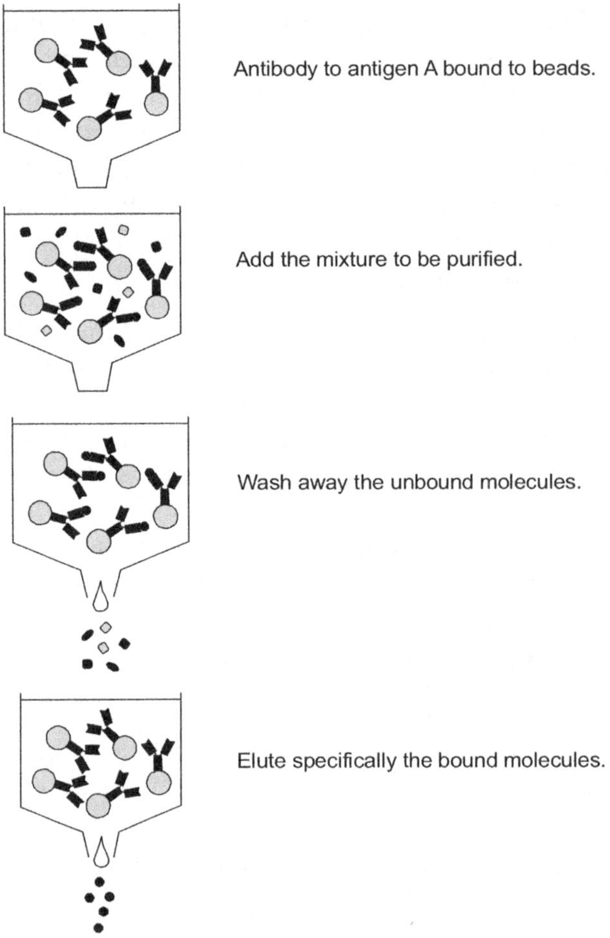
Antibody to antigen A bound to beads.
Add the mixture to be purified.
Wash away the unbound molecules.
Elute specifically the bound molecules.

Figure 6.28 Purification of proteins using immunoabsorbent column (Janeway, C.A. *et al.*, 2001)

antibody to a particular protein is available, it can be used to purify that protein. For example, interferon (IFN) had been previously purified from WBC to only a 1% purity level. D.C. Secher and D.C. Burke first purified interferons, using an anti-IFN in the immunosorbent column (Figure 6.28). When an impure IFN containing many contaminated molecules is passed through anti-IFN immunosorbent column, the IFN binds to anti-IFN and impurities are passed down the column. Later the bound TFN is eluted from anti-IFN. They have achieved several thousandfold increase in purity, in a single passage. Monoclonals immobilized to a cyanogen-bromide-activated chromatographic matrix like Sepharose are valuable for the purification of any protein of medical or research use, or recombinant proteins. By this immunosorbent method, one can achieve 5200-fold purification with up to 97% recovery. Monoclonals are also used as biosensors in affinity purification.

Antibody Chips

In the post-genomic era, antibodies are being utilized for generating antibody chips for the identification of newer human proteins and for the deciphering of protein interaction and protein pathways. Monoclonals against a particular protein can be used to identify its gene. The expression and function of a newly identified gene can be known by raising the monoclonals against the predicted protein that it encodes. It is envisaged that such information can be of significant value in our understanding of normal biological processes and in consequent attenuation of various human diseases and also to determine the abnormal expression of a particular gene or a protein.

COSMETICS

A unique commercial use of monoclonals has been patented by the Gillette Company. They have included an antibody in their deodorant. This antibody blocks the bacterial enzyme necessary for the production of unpleasant-smelling compounds produced by *Staphylococcus haemolyticus*, an organism that causes perspiration in human beings.

As Dr. Richard Siegel, in charge of the pharmaceutical development at Centocor (Malvern, PA), outlined in a presentation at the recent "SRI World Antibody Summit," successful introduction of antibody products has been in the area of therapeutics, with a little application to diagnostics. The complexity of the technology, high cost, and long development time, constrain the search for recombinant diagnostic antibodies, as biotech companies have aggressively pursued blockbuster antibodies for treatment of cancer and autoimmune diseases. It is estimated that more than 700 monoclonals are currently in clinical trials sponsored by over 200 Biotech companies.

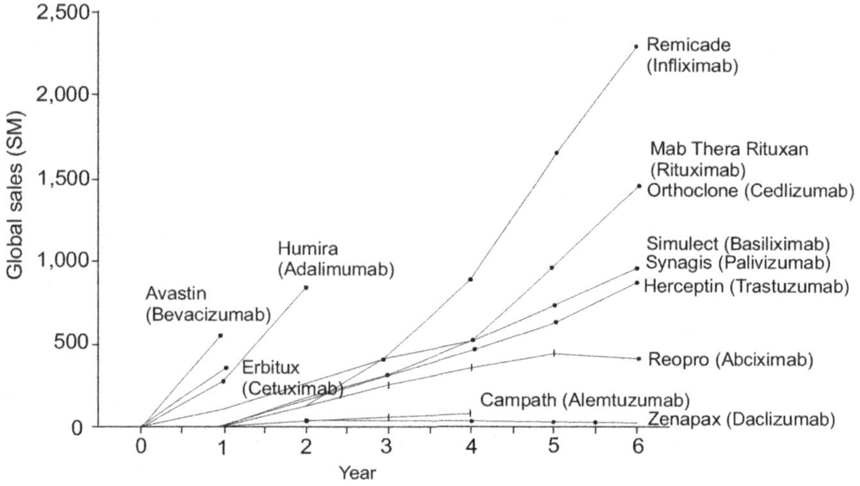

Figure 6.29 Graph illustrating the global sales of monoclonal therapeutics

The monoclonal antibody market is one of the fastest growing and most lucrative sectors of the pharmaceutical industry, with an exceptional 48.1% growth between 2003 and 2004. It has the potential to triple in value over the next six years and reach $30.3 billion in 2010, driven by technological evolution from chimeric and humanized to fully human antibodies. The global sales of monoclonal therapeutics show good sales (Figure 6.29) than any other therapeutics. Rituxan has generated substantial revenues for the companies involved in its development. Bexxar revenue has shown 1.2 million dollars in the last three months of 2003. Zevalin revenue reached $19.6 million during 2003.

Thirty years of research in monoclonals have had a tremendous impact on the biological, medical, pharmaceutical and chemical sciences. Their availability has allowed investigators to ask few questions and to develop new insights and applications that will ultimately benefit human health and the human conditions.

Monoclonals will continue to have important and positive impact on scientific endeavours in the near future. New target antibodies to histones and DNA, cell adhesion molecule or specific enzymes are under early development. We can foresee intense R&D activity in the field of anti-cancer antibody and in the generation of a number of novel therapeutics.

REFERENCES

Abbas, A.K., Lichtman, A.K., Pober, S. (1998). *Cellular & Molecular Immunology.* Harcourt Brace & Company, Asia Pvt. Ltd., WB Saunders Company.

Avidson, I.W. (1992). "Rapid immunoassays." *Anal. Proc.* **29**: 459–460.

Ball, W.J., Kasturi, R., Dey, P., Tabet, M., Donnell, O.S., Hudson, D. and Fishwild, D. (1999). "Isolation and characterization of human monoclonal antibodies to digoxin." *J. Immunol.* **163**: 2291–2298.

Barbara Gatto (2004). "Monoclonal antibodies in cancer therapy." *Current Medi. Chemi. Anticancer Agents.* Vol. **4**: No. 5, 411–414.

Beesley, J.E. (ed.).(1993). *Immunocytochemistry – A Practical Approach.* IRL Press, Oxford.

Burrows, F.J. and Thrope, P.E. (1993). "Eradication of large tumours in mice with an immunotoxin directed against tumour vasculature." *Proc. Natl. Acad. Sci.* USA. **90**: 8996–9000.

Califf, R.M. for the EPIC investigators. (1994). "Use of a monoclonal antibody directed against the platelet glycoprotein IIb/IIa receptor in high risk angioplasty." *N. Engl. J. Med.* **330**: 956–1007.

Chanock, R., Crowe, J., Murphy, B. and Burron, D. (1993). "Human monoclonal antibody Fab fragments cloned from combinatorial libraries – potential usefulness in prevention and/or treatment of major human viral diseases." *Infectious Agents and Diseases.* **2**: 118–131.

Cobbald, S.P. and Waldman, H. (1984). "Therapeutic potential of monoclonal antibodies." *Nature.* **308**: 460–462.

Cone, R.A. and Whaley, K.J. (1994). "Monoclonal antibodies for reproductive health – Part I, Preventing sexual transmission of disease and pregnancy with topically applied antibodies." *Am. J. Reprod. Immunol.* **32**: 114–131.

Corren, J., Ashby, M. and Casale, T.B. (2003). "Omalizumab, a recombinant humanized anti-IgE antibody reduces asthma related emergency room visits and hospitalization in patients with allergic asthma." *J. Allergy Clin. Immunol.* **111**: 87–90.

Datta, P., Hinz, V. and Klee, G. (1996). "Superior specificity of monoclonal based assay for digoxin." *Clin. Biochem.* **29**: 541–547.

Fagerberg, J., Ragnhammar, P., Liljefors, M., Hjelm, A.L., Mellstedt, H. and Frodin, J.E. (1996). "Humoral anti-idiotypic and anti-anti-idiotypic immune response in cancer patients treated with monoclonal antibody 17-1A." *Cancer Immunol. Immunother.* **42**: 81–87.

Feldmann, M., Brennan, F.M. and Maini, R.N. (1996). "The role of cytokine in rheumatoid arthritis." *Ann. Rev. Immunol.* **14:** 397.

Feldmann, M., Elliot, M.J., Woody, J.N. and Maini, R.N. (1997). Anti-tumour necrosis factor –a therapy of rheumatoid arthritis." *Adv. Immunol.* **64**: 283–350.

Fields, B.A., Goldbaum, F.A., Ysern, X., Poljak, R. and Mariuzza, R.A. (1995). "Molecular basis for antigen mimicry by an anti-idiotope." *Nature.* **374**: 739–742.

Ghetie, V. and Vitetta, E. (1994). "Immunotoxins in the therapy of cancer – from bench to clinic." *Pharmac. Ther.* **63**: 209–234.

Glennie, M.J. and Johnson, P.W. (2000). "Clinical trials in antibody therapy." *Immunol. Today.* **21**: 403–410.

Gorman, S., Clark, M.R., Routledge, E.G., Cobbold, S.P. and Waldman, H. (1991). "Reshaping therapeutic CD4 antibody." *Proc. Natl. Acad. Sci.* USA. **88**: 4181–4185.

Gura, T. (2002). "Therapeutic antibodies–Magic bullets hit the target."*Nature.* **417**: 584–586.

Holliger, P. Wing, M., Pound, J.D., Bohlen, H. and Winter, G. (1997). "Retargeting serum immunoglobulin with bispecific diabodies." *Nature Biotechnol.* **15** : 632–636.

Holliger, P., Brissinck, J., Williams, R.L., Thielemans, K. and Winter, G. (1996). "Specific killing of lymphoma cells by cytotoxic T-cells mediated by a bispecific diabody." *Protein Eng.* **9**: 299–305.

Holliger, P., Prospero, T. and Winter, G. (1993). "Diabodies - small bivalent and bispecific antibody fragments." *Proc. Natl. Acad. Sci.* USA. **90**: 6444–6448. http://en.wikipedia.org/wiki/Monoclonal_antibody

Hudson, P.J. and Souriau, C. (2003). "Engineered antibodies." *Nature Medicine.* **9**: 129–134.

Jacques Dantal and Jean Paul Soulillo (2005) . "Immunosupressive drugs and the risk of cancer after organ transplantation." *N. Engl. J. Med.* **352**: 1371–1373.

Janeway, C.A., Travers, P., Walport, M., Shlomchik, M.J. (2001). Immunology 5 – *The Immune System in Health and Disease.* Garland Publishing, New York.

King, D.J. (1998). *Applications and Engineering of Monoclonal Antibodies.* Taylor & Francis Ltd., London.

Kovari, L.C., Momany, C. and Rossman, M.G. (1995). "The use of antibody fragments for crystallization and structure determination." *Structure.* **3**: 1291–1293.

Maloney,D.G., Grillo-Lopez, A.J., White, C.A., Bodkin, D., Schilder, R., Neidhart, J.A., Janakiraman, N., Foon, K.A., Liles, T.M., Dallaire, B.K., Wey, K., Royston, I., Davis, T. and Levy, R. (1997). "IDEC-C2B8 (Rituximab) anti-CD20 monoclonal antibody therapy in patients with relapsed low-grade non-Hodgkin's lymphoma." *Blood.* **90**: 2188–2195.

Marcus Cooke, George, C., Clark, Leo Goeyens and Willy Baeyens. (2000). "Environmental bioanalysis of dioxin." *Todays Chemist At Work.* Vol.**9**, No.7, 15, 16. 19. http://pubs.acs.org/hotartcl/tcaw/00/jul/cooke.html

Marx, J.L. (1989). "Monoclonal antibody and their applications." In: *A Revolution in Biotechnology.* Marx, J.L. (ed.). Cambridge University Press, Cambridge, New York.

Melton, R.G. and Sherwood, R.F. (1996). "Antibody-enzyme conjugates for cancer therapy." *J. Natl. Cancer Inst.* **88**: 153–165.

Michaeli, D. (2005). "Vaccines and monoclonal antibodies." *Semin. Oncol.* 32: (6) 82–86.

Mitchell, E., Reff, Kandhasamy Hariharan and Gay Braslawsky. (2002). "Future of monoclonal antibodies in the treatment of hematologic malignancies." *Cancer Control.* Vol. **9**: No.2., 152–166.

Miyashita, H., Karaki, Y., Kikuchi, M. and Fujii, I. (1993). "Prodrug activation via catalytic antibodies." *Proc. Natl. Acad. Sci.* USA. **90**: 5337–5340.

Neuberger, M., Williams, G.T. and Fox, R.O.(1984). "Recombinant antibodies possessing novel effector functions." *Nature.* **312**: 604–608.

Oettgen, H.F. (1990). Biological agents in cancer therapy–Cytokines, monoclonal antibodies & vaccines." *J. Cancer. Res. Clin. Oncol.* **116**: (1) 116–119.

Peresics, O., Webb, P.A., Holliger, P., Winter, G. and Williams, R.L. (1994). "Crystal structure of a diabody, a bivalent antibody fragment." *Structure* **2**: 1218–1226.

Pollack, S.J., Jacobes, J.W. and Schulz, P.G. (1996). "Selective chemical catalysis by an antibody." *Science.* **234**: 1570–1573.

Quesniaux, V.F. (1991). "Monoclonal antibody technology for cyclosporine monitoring." *Clin. Biochem.* **24**: 37–42.

Raag, R. and Whitlow, M. (1995) "Single-chain Fvs." *FASEB J.* **9**: 73–80.

Rees, A.R., Staunton, D., Webster, D.M., Searle, S.J., Henry, A.H. and Pedersen, J.T. (1994). "Antibody design – beyond the natural limits." *Trends Biotechnol.* **12**: 199–206.

Reichert, J. and Pavolu, A. (2004). "Monoclonal antibodies market." *Nat. Rev. Drug Discov.* **3**: 383–384.

Reichert, J.M. (2001). "Monoclonal antibodies in the clinic." *Nat. Biotechnol.* **19:** 819–822.

Reichert, J.M. (2002). "Therapeutic monoclonal antibodies- trends in development and approval in the US." *Curr. Opin. Mol. Ther.* **4**: 110–118.

Riechmann, L., Weill, M. and Cavanagh, J. (1992). "Improving the antigen affinity of an antibody Fv fragment by protein design." *J. Mol. Biol.* **224**: 913–918.

Roitt, I. Brostoff, J., Male, D. (2000). *Immunology,* 6th edn. Mosby, Edinburgh.

Senter, P.D. and Svennson, H.P. (1992). "A summary of monoclonal antibody-enzyme/prodrugs by antibody targeted enzyme conjugates for cancer therapy." *Cancer Res.* **52**: 4484–4491.

Sharman Gold, R. (1997). "Monoclonal antibodies–the evolution from '80 magic bullets to mature, mainstream applications as clinical therapeutics." *Genetic Engineering News.* 17. August 4, 35.

Talwar, G.P., Raghupathy, R., Gupta, S.K. and Bal, V. (2004). "Immunotherapy". In: *Concepts in Biotechnology.* (ed.). by Balasubramanian *et al.* Universities Press (India) Pvt. Ltd., Hyderabad.

Thakur, I.S. (1996). "Use of monoclonal antibodies against dibenzo-*p*-dioxin degrading *Sphingomonas* sp. strain RW1." *Appl. Microbiol.* **22**(20): 141–144.

The PREVENT Study Group. "Reduction of respiratory syncytial virus hospitalization among premature infants and infants with bronchopulmonary dysplasia using respiratory syncytial virus immunoglobulin prophylaxis." *Pediatrics.* **99**: 93–99.

Tramontano, A., Janda, K.D. and Lerner, R.A. (1986). "Catalytic antibodies." *Science.* **234**: 1566–1570.

Vaughan, T., Osbourn, K. and Tempest, P.R. (1998). "Human antibodies by design." *Nat. Biotechnol.* **16**: 535–539.

Vietetta, E.S. and Uhr, J.W. (1994). "Monoclonal antibodies as agonists–an expanded role for their use in cancer therapy." *Cancer Res.* **54**: 5301–5309.

Waldman, T.A. (2003). "Immunotherapy – Past, present & future." *Nature Medicine.* **9**: 269–277.

Wentworth, P., Datta, A., Blakey, D., Boyle, T., Partridge, L.J. and Blackburn, G.M. (1996). "Towards antibody-directed 'abzyme' prodrug therapy, ADEPT – carbamate prodrug activation by a catalytic antibody and its application *in vitro* to human tumour cell killing. "*Proc. Natl. Acad. Sci.* USA. **93**: 799–803.

7 MONOCLONAL ANTIBODIES AS HOPEFUL DRUGS

INTRODUCTION

"The unbridled optimism that surrounded monoclonal antibodies in the 1980s was infectious. You had to be the world's toughest cynic not to be dazzled. Got cancer? No problem. Like heat-seeking missiles, monoclonal antibodies tipped with poisons or radioactive isotopes will home in on malignant cells and will deliver their deadly payloads, wiping out cancer while leaving the normal cells intact. How about an infectious disease? All would be well. Monoclonals surround the marauding viruses and bacteria, like goombahs from Tony Soprano's crew, muscling them into secluded byways where killer cells of the immune system will provide them an offer they couldn't refuse," says Donald L. Drakeman, the President and CEO of monoclonal maker Medarex in Princeton, N.J. He says further that the antibodies are simply easier to develop than the traditional drugs that are composed of small, inorganic molecules. The development of a drug takes at least 16 years from the time of its discovery (Figure 7.1).

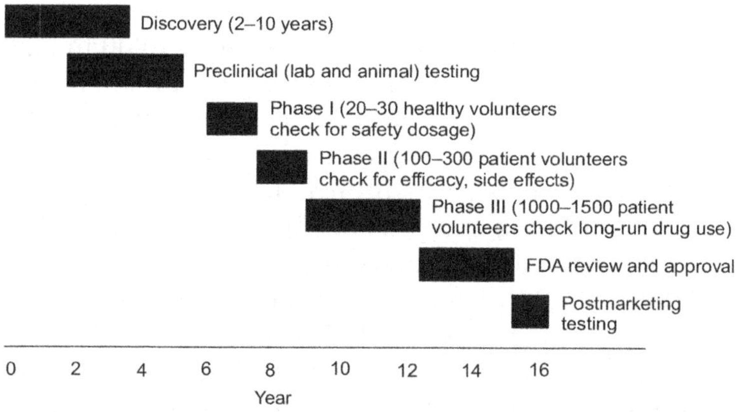

Figure 7.1 Time taken for various stages of drug development

As monoclonals are large molecules and might not be suitable for every disease, they take only one or two years to come up for testing, in contrast to small-molecule drugs which take five years. That speed translates into savings: it costs only $2 million to make ready a monoclonal antibody for clinical testing, Drakeman estimates, compared with $20 million for a traditional drug. And despite the FDA's hesitancy to approve Genentech's asthma therapy, he states that monoclonals have so far had a higher success rate than small-molecule drugs in clearing regulatory hurdles. "Antibodies are almost never toxic," he explains. Ironically, monoclonals might be victims of their own success: market analysts are predicting that companies won't have sufficient production facilities to make them all. The biotechnology industry has anticipated this problem.

THERAPEUTIC MONOCLONALS

MONOCLONALS AGAINST INFECTIOUS DISEASES

Although immunotherapy has been used for more than hundred years, it became less important when antimicrobial agents came into widespread use. In the 1970s investigators began to re-examine immunotherapy for potential use in gram-negative infections. Polyclonal antiserum against J5 mutant of *E. coli* has been shown to be effective in treating patients with bacteraemia and septic shock.

Bacteraemia

E5 MONOCLONAL

Bacteraemia, a blood infection, accounts for 24,000 to 72,000 deaths annually in US hospitals because the most common bacterial strains are resistant to antibiotic treatment. The toxicity is due to lipid A of lipopolysaccharide, an endotoxin that triggers a set of excessive immunological responses resulting in fever, cell lysis, increased heart beat, hyperventilation and possibly fatal organ failure. One approach to control gram-negative bacteraemia has been to use antibodies against the endotoxin. Both the human and mouse monoclonals of IgM type, called E5, bind to endotoxin. When tested in clinical trials, they were found to control the disease. It reduced the mortality and improved the outcome of multiorgan failure in patients of refractory shock. No adverse side effects were observed with human monoclonals. The mouse monoclonal produced only a minor immunological response in a few patients. Further evaluation of E5 antibody is warranted in the treatment of patients with neutropenia, burns, and shock.

Initial economic estimates put the cost per dose at $ 3750 or $ 2.3 billion per year, if all patients with bacteraemia were treated with immunotherapeutic monoclonals. Of course, this estimated cost must be balanced against the expenditures of treating bacteraemia by current practices and the fact that if these antibodies are truly effective, the lives may be saved.

Respiratory Syncytial Viral Infections

Palivizumab (Synagis)

Palivizumab is the first successfully developed monoclonal antibody to combat an infectious disease caused by the respiratory syncytial virus (RSV)—a ubiquitous and highly contagious virus. Approximately two-thirds of infants are infected with RSV during their first year of life and almost 100% by the age of two (Figure 7.2). Seasonal outbreaks occur each year in winter. Most of these RSV infections cause minor upper respiratory tract illness and cold-like symptoms. It has been estimated that these viral infections cause up to 125,000 hospitalizations annually in children during their first year of life and that between 50–80% of US communities show bronchiolitis hospitalizations from November through April. Infants with chronic bronchopulmonary dysplasia (BPD), infants with a history of premature birth (≤35 weeks gestational age), and children with haemodynamically significant coronary heart disease and those with certain form of heart diseases are at high risk. About 1–2% of these infants die.

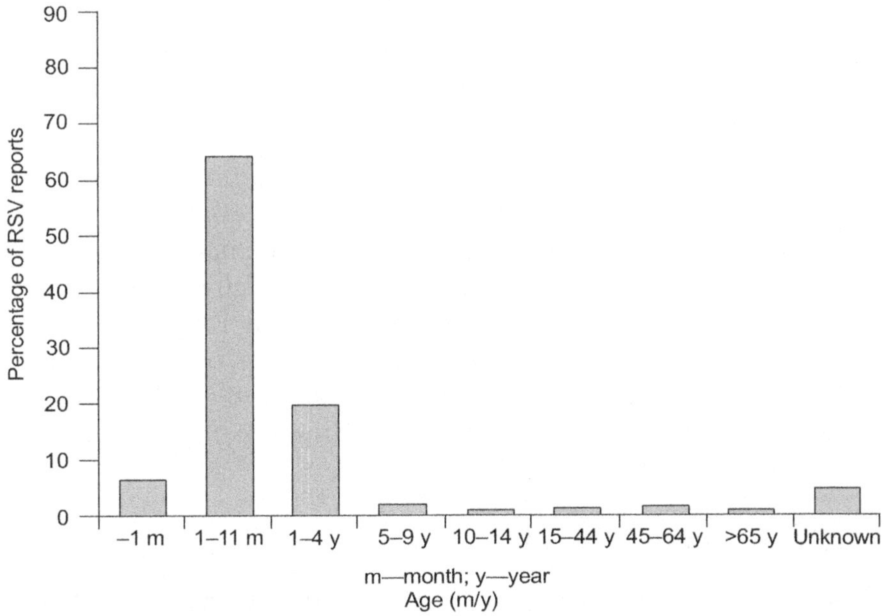

Figure 7.2 Graph illustrating the prevalence of RSV infection (Archive, 2004-News.Vol.16. No.1).

Palivizumab is a humanized monoclonal obtained by grafting 6 human CDRs in the murine monoclonal 1129, using recombinant DNA technology. It is approved as a prophylactic drug or vaccine. It binds to the RSVF glycoprotein, inhibits the fusion activity of virus and also exhibits neutralization. A single intramuscular injection of this drug reduces 45% hospitalization (Figure 7.3). It is given to premature babies once a month for six months.

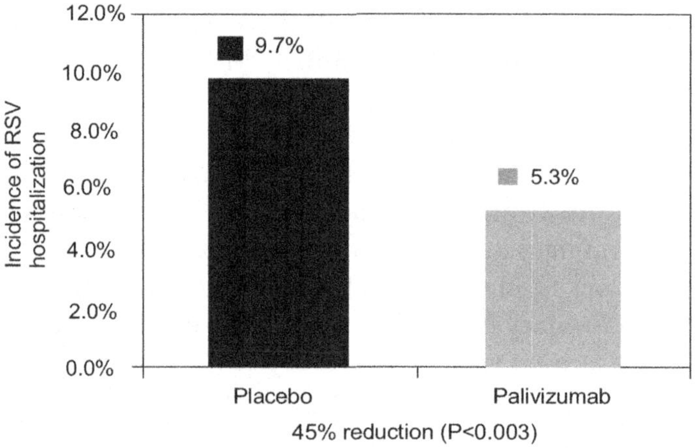

Figure 7.3 Graph showing 45% reduction of hospitalization of infants after the vaccination with Palivizumab (www.synagis.com/.../heart disease.a spx.)

Along with its desired effects, a medicine may cause some unwanted effects. The more common side effects are the bluish colouring of fingernails, lips, skin, palms, or nail beds (in patients with heart disease), breathing difficulties, ringing or buzzing in the ears, skin rashes, etc. The less common or rare side effects are abdominal pain, diarrhoea, dizziness, fast-, slow-, or irregular heartbeat, loss of appetite, lump in abdomen, nausea and weakness. Although not all of these side effects may occur, if they do occur, they may need medical attention.

West Nile Viral Infection

West Nile virus alarmed the Americans when it made its first appearance in New York city in 1999. Since then it has spread from coast to coast, sickened more than 16,000 Americans and killed more than 600. As the virus spreads, medical investigators hastened their research to develop an effective vaccine or therapy. None currently exist, but a newly published paper by researchers at Washington University in St. Louis points out a promising treatment.

The scientists of Washington University made 46 monoclonal antibodies against the E-protein antigen of West Nile virus and then eliminated the less effective ones through a tedious molecular-level screening process. They then turned to Rockville, Maryland-based MacroGenics Inc., to create a human-like version of the most effective antibody. Macrogenics tailored the part of the antibody that cripples the West Nile virus into the scaffold of a human antibody. The monoclonal antibody was several hundred times more potent in cell culture tests than the antibodies obtained from people who had recovered from this viral infection. A second round of tests in mice confirmed that the new antibodies retained their ability to prevent West Nile virus.

Mucosal Infections

A large percentage of death is due to mucosal infections. Even though systemic applications of monoclonals have often been a routine procedure, topical passive immunization with monoclonals may offer a new opportunity for preventing these infections. Recent advances like combinatorial libraries, and transgenic antibodies from animals and plants, allow development of a new era of mucosal monoclonal-based products.

From the seropositive persons using combinatorial libraries, fully human monoclonals against human pathogens can be generated. For example, from a single bone marrow donor, human monoclonals were prepared against HIV, respiratory syncytial virus (RSV), cytomegalovirus, herpes simplex virus types 1 and 2, varicella zoster virus, and rubella virus. Monoclonals against these viruses can even be obtained from naive libraries prepared from unexposed persons if the library has a large-enough repertoire. Therefore, antibodies against the pathogens that are lethal to humans can be generated.

Antibodies can be delivered to the mucus either as gel, solution and spray, or by a polymer, so that they can trap the infective organisms in the mucus (Figure 7.4). This new interest in mucosal antibodies may be partially due to the increasing recognition of the importance of mucosal immunity.

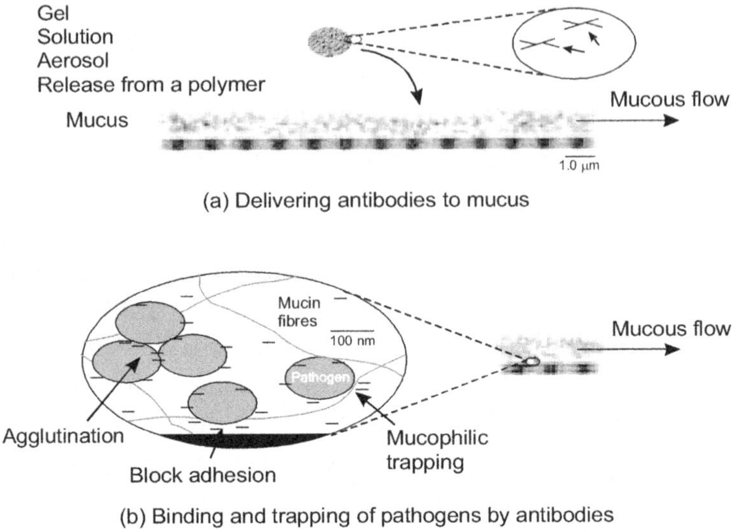

Figure 7.4 Mechanism of topically applied mucosal antibodies (Zeitlin, L., *et al.*, 1999)

Only two clinical trials have evaluated the efficiency of topically applied monoclonals: intranasally delivered anti-RSV in infants at high risk and orally delivered monoclonals against *S. mutans* in adults. Topically applied monoclonals have certain advantages:

- Show less immunogenicity with no major adverse side effects.
- Provide immediate protection against infection.
- Have minimal interaction with circulating immune cells.
- Maintain the integrity of mucosal surfaces since sIgA cannot activate complement by the classic pathway.
- Have less drug resistance problem since the pathogen will not replicate and evolve in the presence of antibody on mucosal surface.
- Even if antibody-resistant strains are produced, a new monoclonal directed towards the mutated epitope can be developed within a short period of 1–3 months.

From a public health perspective, monoclonals are the most promising for preventing gastrointestinal, respiratory, and reproductive tract infections. These infections cause almost 11 million deaths annually worldwide, accounting for more than 50% of the death caused by communicable diseases and 22% of death by all other causes. Sexually transmitted diseases (STDs) accounted for 87% of all cases reported among the top ten most frequently reported diseases in 1995 in the United States; more than 12 million Americans are infected with STDs each year at an estimated annual cost of more than $12 billion. Personal protection provided by over-the-counter antibody-based technology could play a similar role in future emerging epidemic diseases.

Diarrhoeal diseases Orally delivered bovine antibodies were 100% effective in preventing the infections of rotavirus, enterogenic *E. coli*, *Shigella* and also in preventing necrotizing enterocolitis in animal models and in humans. As diarrhoeal diseases are more prevalent and endemic in developing countries, monoclonals could be delivered orally as a supplement with food or water in a more inexpensive way. Significant levels of functional antibody survive after treatment with pepsin at pH 2 or with a pool of pancreatic enzymes at pH 7.5 *in vitro*, without any degradation by these proteolytic enzymes. Ingested IgA in milk in infants, or intact antibody delivered orally with an antacid survives the passage through the gastrointestinal tract. As monoclonals pass through the gastrointestinal tract with great speed, the antibodies should be delivered more than once a day.

Respiratory diseases Topically applied or nasally delivered antibodies were 100 times more effective than systemic delivery. Anti-RSV monoclonals (MEDI-493) have shown to be 100 times more effective than an equal quantity of a polyclonal preparation. These findings suggest that 10,000 times less anti-RSV monoclonals would be required for topical applications than for systemically delivered polyclonal preparations. Protective systemic doses of MEDI-493 are approximately 100 mg (15 mg/kg), so <1 mg topically applied might suffice for protection. The residence half-life of a nasally delivered monoclonal is a little less than one day, suggesting that applying once a day in the form of nasal drops or by the aerosol may deliver severalfold protection even during the influenza season.

STDs Monoclonals have shown to be protective against transmission of *C. albicans*, *C. trachomatis*, HSV, HIV, and *T. pallidum* in animal models, either by the experimental delivery of monoclonals to the vagina in solution, gels, etc., or by the sustained release devices for long-term delivery. No significant inactivation of monoclonals occurred over the pH range (pH 4 to 7) of the human vagina for at least 24 hours at 37°C and were found to be stable when they are stored in seminal fluid or cervical mucus for 48 hours at 37°C. A single vaginal application may provide protection for at least 1 day or probably several days when the user forgets to apply the monoclonals. Until effective and safe vaccines are developed, vaginal delivery of a cocktail of anti-STD-pathogen monoclonals might be an effective new method for broad-spectrum protection against STDs.

AIDS Higher affinity or higher neutralization activity of monoclonals can be obtained by site-directed mutagenesis. A 420-fold increase in the affinity of anti-HIV1-Fab monoclonal is shown by **CDR walking**, in which CDRs are mutated either in turn or in parallel. These mutated monoclonals neutralize more HIV strains than the original monoclonals. Further an sIgA (dimer) or IgM (pentamer) monoclonal can dramatically enhance the potency of an antibody by increasing the avidity. Anti-*E. coli*-IgM was found to be 1000-fold more effective in protecting neonatal rats than its class-switched IgG (both *in vitro* and *in vivo*). A 1000-fold increase in avidity could translate into a 1000-fold decrease in dose and subsequent cost. A large dose of a highly potent monoclonal can also substantially increase the duration of protection.

MONOCLONALS AGAINST CANCER

When a normal cell is transformed to a cancerous cell, it usually begins a high-level expression of several genes that were previously either unexpressed or expressed at extremely low levels. Some of the resultant proteins are found on the cell surface and are termed tumour surface antigens (TSAs). TSAs potentially represent unique surface antigens, and antibodies raised against specific TSAs are likely to be approved for oncological applications.

Monoclonal antibody therapy is an effective way of specifically targeting certain kinds of cancer cells, with a low degree of toxicity to normal cells. Monoclonals are intravenously administered to patients, generally on an outpatient basis. Once in the bloodstream, monoclonal antibodies travel throughout the body and attach themselves to the cells that have the specific target antigen, such as cancer cells. This alerts the body's own immune system to recognize the cancer cell and helps to destroy it. Although some normal cells that also have the specific target antigen may be affected along with the cancer cells, the body can usually replace these normal cells in the course of treatment.

There are a number of considerations in using monoclonals for cancer therapy. While selecting the target antigen, it is of importance that the antigen must be unique to the cancer cell. The immunogenicity of a monoclonal, serum half-life, cost and availability of it must be of prime consideration. It has to be decided whether to use these monoclonals alone as naked antibodies, or as conjugated antibodies with radioisotopes, toxin or cancer drug.

The FDA-approved and commercially available first-generation monoclonals are the following.

Rituximab (Rituxan)

The first monoclonal approved by FDA in 1996 for the treatment of non-Hodgkin's lymphoma (a cancer of B cell). It is a chimeric or humanized (95%), unconjugated monoclonal directed against the CD20 antigen, a signature B cell antigen.

CD20 antigen is found on the surface of all normal and abnormal B cells that are found in some of the most common types of non-Hodgkin's lymphoma. This antigen arises during the pre-B-cell stage of B-cell differentiation. It is a good cancer marker gene since it does not circulate in the plasma as a free protein. It can competitively inhibit the antibody binding to lymphoma cells and can exist on the surface of $CD20^+$ cells even after antibody binding. It cannot be internalized, but subsequently down-regulated on antibody binding.

CD20 is a transmembrane protein, playing a major role in activation, proliferation and differentiation of B cells. The cytoplasmic anchor contains the phosphorylation sequences for protein kinase C, calmodulin and casein kinase 2. When Rituximab binds to CD20, it induces apoptosis through the calcium influx within the cells (Figure 7.5). An increase in the intracellular calcium activates the Src family tyrosine kinases, resulting in further phosphorylation of inner cytoplasmic CD20 and phospholipase Cg. At the same time, there is an upregulation of *c-myc* and *b-myb* mRNA, an increase in adhesion molecule expression and an upregulation of MHC class II proteins. The ultimate result is caspase-3 activation, that causes cell apoptosis or programmed cell death. It blocks the growth factor receptor effectively, thereby arresting proliferation of tumour cells.

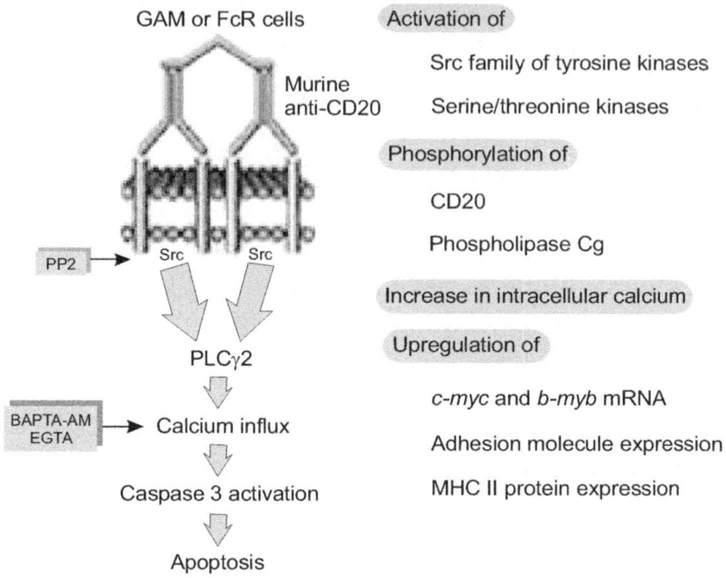

Figure 7.5 Effect of CD20 cross-linking (Shan, D. *et al.*, 1998)

When Rituximab locks on to the CD20 antigen on the surface of a B cell, the cell may be destroyed directly through apoptosis and also the body's natural defences are alerted. The Fc fragment of an IgG monoclonal has the following 2 roles.

1. It triggers ADCC (antibody dependent cell-mediated cytotoxicity) or phagocytosis by activating the Fc gamma receptors of cells like monocytes, macrophages and natural killer cells (Figure 7.6a) which then secrete cytokines that attack the target cell.
2. It triggers complement-dependent cell lysis by activating the C1q protein (a member of the protein enzymatic pathway called complement), which then activates C1r, which in turn produces C1 esterase. This esterase activates C2 and C4, and hence a cascade of activations (chemokines, anaphylatoxins) which lead to the polymerization of the C9 fraction of complement into a tubular structure that drills a hole (Figure 7.6b) in the lymphocyte membrane.

The macrophage opsonizes the bound tumour cells and also brings about anti-idiotypic antibody formation. The normal B cells also may be destroyed when Rituximab is used but soon new B cells are recovered from stem cells. However, the stem cells and pro- or pre-B-cells in the bone marrow that develop into functional B cells do not have CD20 on their surface.

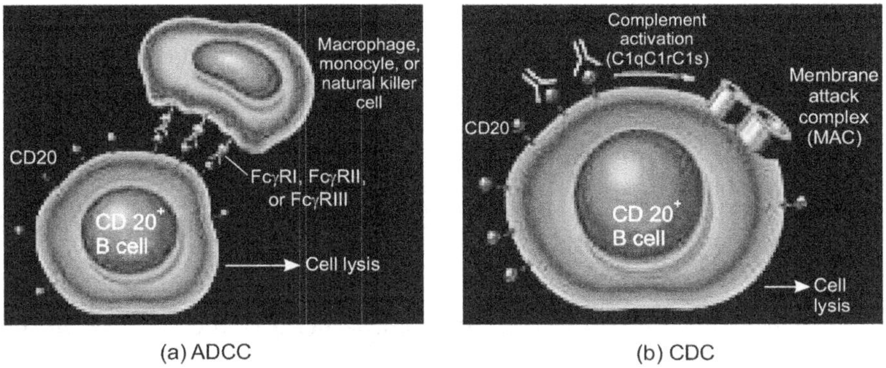

Figure 7.6 Mechanism of action of Rituximab on tumour cell

The efficacy and safety of Rituximab as a single-agent therapy has been reported in patients with low-grade or follicular NHL in four phase I/II studies. In a phase III study, adverse events associated with treatment have not been severe and were primarily classified as grade 1 or grade 2. They are not due to the drug itself and are infusion related, and included transient fever, chills, nausea, and headache. Only one patient (< 1%) developed human anti-chimeric antibodies (HACA). The overall response rate has been 48% for the intent-to-treat population (randomized control trial patients) and 50% for the assessable population (those patients who were randomized took at least one dose of study medicine in both crossover period), with a median time to progression and duration of response in assessable responders of 13.2 and 11.6 months respectively.

Additionally, treatment trials with the combination of Rituximab and CHOP [cyclophosphamide, hydroxydaunoribin (doxorubicin), vincristine (oncovin) and prednisolone/prednisone]; Rituximab and IDEC-Y2B8; and Rituximab and interferon have been investigated in patients with intermediate- and high-grade lymphoma. Figure 7.7 shows the phase III trial of Rituximab/CHOP combination therapy. It increases the overall survival and event-free survival of the patients comparable to those receiving CHOP alone. Rituximab can be given to non-Hodgkin lymphoma patients with a dose of 375 mg/m^2 on a weekly basis for 4 weeks and then it is reevaluated after two weeks.

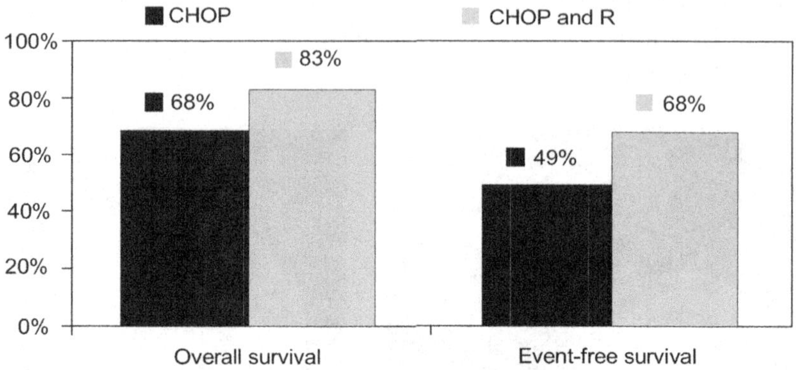

Figure 7.7 Comparison of CHOP and Rituximab, and CHOP treatments in the survival of lymphoma patients (Czuczman, M.S., 1999).

Ocrelizumab is the humanized version of this monoclonal in clinical trials at the moment. Also a fully human anti-CD20, HuMaxCD20, is also in clinical trials.

Rituxan has caused severe fatal infusion reactions within 30 to 120 minutes of first infusion. Rituxan administration to patients who develop severe infusion reactions should be discontinued and be given proper medical treatment. The signs and symptoms of severe infusion reactions may include urticaria, hypotension, angioedema, hypoxia, or bronchospasm, myocardial infarction, ventricular fibrillation, or cardiogenic shock. Acute renal failure requiring dialysis with instances of fatal outcome has been reported in the setting of TLS (tumour lysis syndrome) following the treatment of non-Hodgkin's lymphoma (NHL) patients with Rituxan. Severe mucocutaneous reactions, some with fatal outcome, have also been reported in association with Rituxan treatment.

Trastuzumab (Herceptin)

About 20 to 30% of breast cancer may be due to the overexpression of a protooncogene known as the HER2 gene (also referred to as c-erbB-2 HER2/neu). Under normal conditions, the two copies of the HER2 gene present in a cell produce small amounts of a human epidermal growth factor receptor known as the HER2 protein. This plays a role in the transmission of signals that ensure controlled cell growth with a regulated rate of division. Overexpression of HER2 protein leads to a state of sustained activation in which growth-promoting signals are transmitted to the cell nucleus on a permanent basis, resulting in an uncontrolled growth and an increased rate of division. This overexpression is generally due to the amplification of the HER2 gene (Figure 7.8a). This phenomenon is observed not only in breast cancer, but also in tumours of the lung, intestine, prostate, fallopian tubes, and other parts of the body.

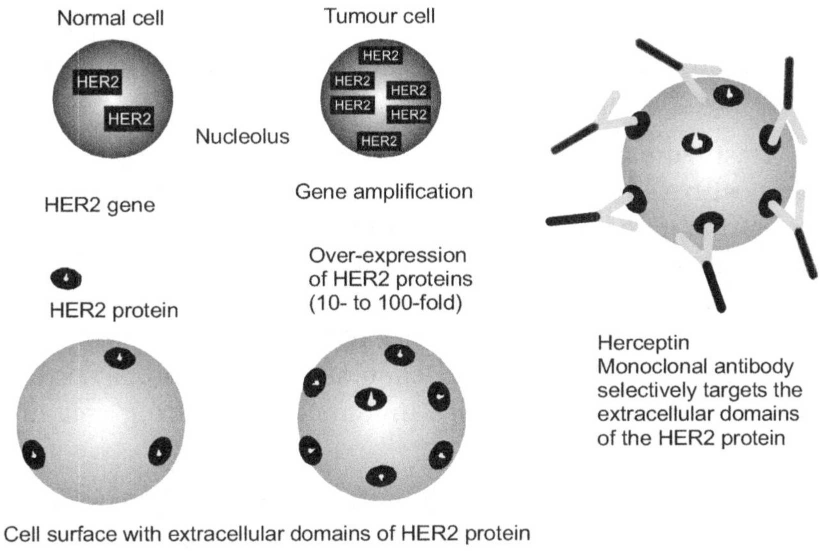

(a)

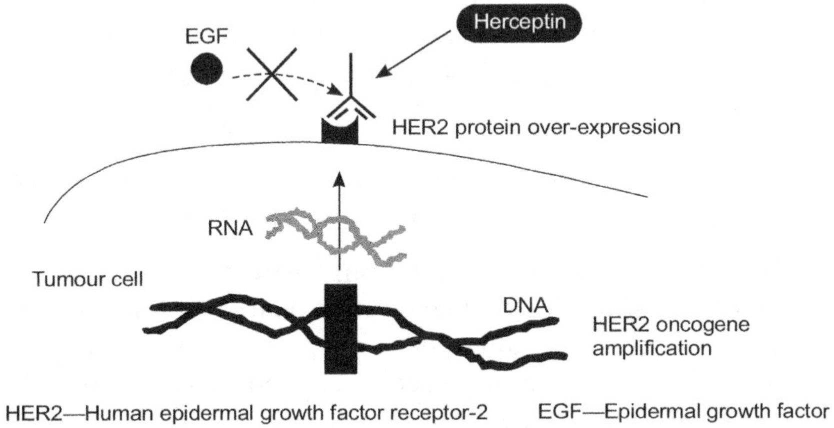

(b)

Figure 7.8 (a) HER2 gene and its protein in normal and tumour cells (b) Action of herceptin on tumour cells (www.roche.com/pages/facets/9/herc1.jpg)

Herceptin, a biological smart missile, reacts only with the cancer cells that have an increased number of HER2 receptors on their surface. It has been approved by FDA in 1998 in the US and in 2000 in Europe. The specific binding of this chimeric antibody to the outward-facing part of the HER2 receptor prevents the transmission of signal to the cell nucleus, downregulates the overexpression of HER2 proteins, their receptors and angiogenesis factors, inhibits the proliferation of tumour cells and also enhances the immune recruitment and ADCC against the tumour cells (Figure 7.8b). Thus, the immune system is stimulated to destroy the cancer cells.

Trastuzumab has achieved notable results in the treatment of HER2/neu-positive advanced metastatic breast cancer and is under extensive evaluation in major clinical trials for its potential efficacy when used in earlier stages of breast cancer. Patients with early stage HER2-positive breast cancer who received Herceptin plus chemotherapy experienced a 52 per cent reduction in the risk of disease recurrence compared to those patients who received chemotherapy alone, according to an interim analysis of two studies announced by the drug's manufacturer, Genentech Inc.

After four years in the study, only 15 per cent of the women treated with Herceptin plus chemotherapy experienced disease recurrence, compared to 33 per cent of women treated with chemotherapy alone. Preliminary survival data showed a 49 per cent improvement in overall survival or a hazard ratio of 0.67, which is equivalent to a 33 per cent reduction in the risk of death. Survival data continue to increase. Herceptin slashed the cancer recurrence by half. "Many in the breast cancer community feel that this is one of the most important breast cancer advances in decades of therapy," says Dr. Sandra J. Horning, President, American Society of Clinical Oncology (ASCO).

Trastuzumab administration can result in the development of ventricular dysfunction and congestive heart failure. Left ventricular function should be evaluated in all patients prior to and during treatment with Herceptin. Discontinuation of Herceptin treatment should be strongly considered in patients who develop a clinically significant decrease in left ventricular function. The incidence and severity of cardiac dysfunction were particularly high in patients who received Herceptin in combination with anthracyclines and cyclophosphamide.

Herceptin administration can result in severe hypersensitivity reactions (including anaphylaxis), infusion reactions, and pulmonary events. Rarely, these have been fatal. In most of the cases, symptoms occurred during or within 24 hours of administration of Herceptin. Herceptin infusion should be interrupted for patients experiencing dyspnea or clinically significant hypotension. Patients should be monitored until signs and symptoms completely resolve. Discontinuation of Herceptin treatment should be strongly considered for patients who develop these symptoms.

Gemtuzumab ozogamicin (Mylotarg)

It is a recombinant humanized (98.3%) IgG4 kappa antibody, conjugated with the cytotoxic anti-tumour antibiotic calicheamicin isolated from fermentation of a bacterium, *Micromonospora echinospora* ssp. *calichensis*. It is given for the treatment of acute myeloblastic anaemia (AML), a cancer of myeloblastic cells. The Fab portion of the antibody binds to CD33 antigen—a sialic-acid-dependent adhesion protein found on the surface of leukaemic blast cells and immature normal cells of myelomonocytic lineage, but not on normal haematopoietic stem cells. 50% of the antibody is loaded with 4–6 moles of the drug/molecule. It has been approved by FDA in 2002 in US and in 2004 in Europe.

Mylotarg is directed against the CD33 antigen expressed by the haematopoietic cells and leukaemic blast cells. The binding of Mylotarg with the CD33 antigen results in the formation of a complex, which is internalized. Upon internalization, the calicheamicin derivative is released inside the lysosomes of the myeloid cell. The released calicheamicin derivative binds to the DNA in the minor groove resulting in DNA double strand breaks and cell death.

Gemtuzumab ozogamicin is cytotoxic to the CD33-positive HL-60 human leukaemia cell line. It inhibits significantly the colony formation in cultures of adult leukaemic bone marrow cells. The cytotoxic effect on normal myeloid precursors leads to substantial myelosuppression, but this is reversible because pluripotent haematopoietic stem cells are spared. In the preclinical animal studies, Gemtuzumab Ozogamicin demonstrates antitumour effects in the HL-60 human promyelocytic leukaemia xenograft tumour in athymic mice.

The efficacy and safety of Mylotarg as a single agent was evaluated in 142 patients in three single-arm open-label studies in patients with

CD33-positive AML in first relapse. The studies included 65, 40, and 37 patients. In studies 1 and 2, patients ≥18 years of age had a first remission duration of at least 6 months. In study 3 (Table 7.1), only patients ≥60 years of age were enrolled and their first remission duration had lasted for at least 3 months. The patients with secondary leukaemia or with white blood cell (WBC) counts ≥30,000/μL were excluded. Some patients were leucoreduced with hydroxyurea or leukapheresis to lower the WBC counts below 30,000/μL in order to minimize the risk of tumour lysis syndrome.

Table 7.1 Percentage of patients by remission category

Type of remission	Study I	Study II	Study III	All studies
Age	≥ 18 years	≥ 18 years	≥ 60 years	
Number taken for studies	$n = 65$	$n = 40$	$n = 37$	$n = 142$
Complete remission (CR)	17	20	11	16
Complete remission to Patient's satisfaction (CRP)	15	13	11	13
Overall response (CR + CRP)	32	33	22	29

The treatment course has been included two 9 mg/m^2 doses separated by 14 days and a 28-day follow-up after the last dose. Although smaller doses have elicited responses in earlier studies, this additional dosage was chosen because it was expected to saturate all the CD33 sites regardless of leukaemic burden. The primary end point of these three clinical studies has been the rate of complete remission where there is a complete absence of leukaemic blast cell in the peripheral blood.

The common side effects of this monoclonal include

i. decrease in bone marrow production,

ii. low red blood cell counts (anaemia),

iii. low blood platelets, infection, bleeding and transfusions, swelling of the membrane inside the mouth,

v. liver problems and skin rashes.

If these symptoms occur, treatment with Mylotarg should be discontinued. Other medicines can be prescribed for chills, fever, nausea, vomiting, headache, changes in blood pressure and low levels of oxygen in the body.

Alemtuzumab (Campath)

Chronic lymphocytic leukaemia (CLL) is the most prevalent form of leukaemia in adults and affects approximately 120,000 patients in the United States and Europe. B-CLL is characterized by an accumulation of leukaemic lymphocytes in the bone marrow, blood, and other body tissues. This accumulation leads to dysfunction of the bone marrow and enlargement of the lymph nodes, liver, and spleen. The symptoms of the disease include fatigue, bone pain, night sweats, decreased appetite, and weight loss.

Campath (alemtuzumab) is used for the treatment of B-cell chronic lymphocytic leukaemia (BCCLL) in the patients who have been treated with alkylating agents, and who have failed fludarabine therapy. It works in an entirely different way than chemotherapy, and it is the first and only drug of this type approved by the FDA in 2001, for the treatment of patients with B-CLL. A recombinant-DNA-derived IgG-1 kappa humanized Campath is targeted against CD52, an antigen expressed by the eosinophil and monocytes, and is used as a clean-up agent.

Campath binds to the CD52 antigen present on the surface of the malignant lymphocytes and induces antibody-dependent lysis or killing. This causes the removal of malignant lymphocytes from the blood, bone marrow, and other affected organs. Campath is used subcutaneously along with Rituximab for patients with relapsed or refractory, low-grade or follicular, CD20-positive, B-cell non-Hodgkin's lymphoma.

Ibritumomab tiuxetan (Zevalin)

It is a murine monoclonal antibody conjugated with a radioactive isotope, used for the treatment of patients with relapsed or refractory, low-grade or follicular, non-Hodgkin's lymphoma. Like Rituximab, it induces apoptosis in $CD20^+$ B-cell lines *in vitro*. The complementarity-determining regions of Ibritumomab bind to the CD20 antigen on B lymphocytes. The chelate tiuxetan, which tightly binds the ^{111}indium or ^{90}Y, is covalently linked to the amino groups of exposed lysines and arginines contained within the antibody. The beta emission from ^{90}Y induces cellular damage by forming free radicals in the target and neighbouring cells.

The Zevalin therapeutic regimen is administered in two steps.

Step 1 It includes one infusion of Rituximab preceding an injection of ^{111}Indium Zevalin (Figure 7.9).

Step 2 It follows step 1 by seven to nine days and consists of a second infusion of Rituximab followed by ^{90}Y-Zevalin.

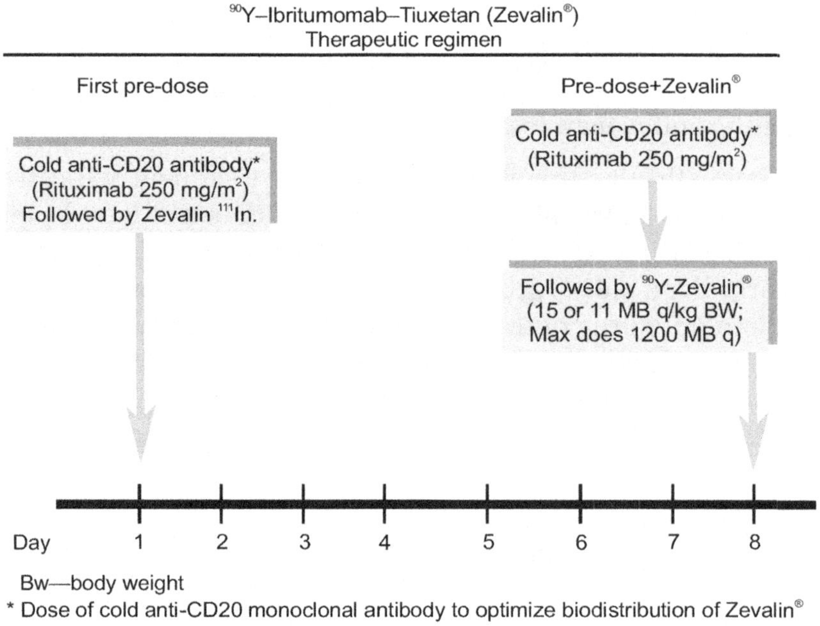

Figure 7.9 Zevalin therapeutic regimen (www2.alasbimnjournal.cl/AlasbimnImages/aj29-a...)

A randomized multicentre clinical study compared the efficacy of the Zevalin therapeutic regimen versus Rituximab 375 mg/m^2 weekly over 4 weeks in 143 patients with relapsed or refractory, low-grade or follicular, transformed B-cell NHL. The patients were considered refractory if they did not achieve a complete or partial response to their last chemotherapy, or if they had disease progression within 6 months for the Zevalin therapeutic regimen arm or within 3 months for the Rituximab arm. Responses were determined using International Workshop response criteria. The overall response rate, in the randomized 73 patients who received the Zevalin therapeutic regimen, was 80% (with 30% complete responses). This is statistically higher than the 56% overall response rate (with 16% complete responses) observed in the 70 patients who received Rituximab (P = .002). The unconfirmed complete response rate was 4% for both the arms.

Figure 7.10 shows the imaging of a 51-year-old male with relapse or refractory, low-grade, non-Hodgkin's lymphoma. In the Cancer Center treatment room, the patient was infused with rituxan (250 mg/m^2 or 465 mg) over a four-hour period. Once the infusion of Rituxan was complete, the Nuclear Medicine Technologist infused 5.0 mCi of ^{111}In-Zevalin over a 10-minute period (the infusion included a 0.22 micron filter attached at the end of the syringe).

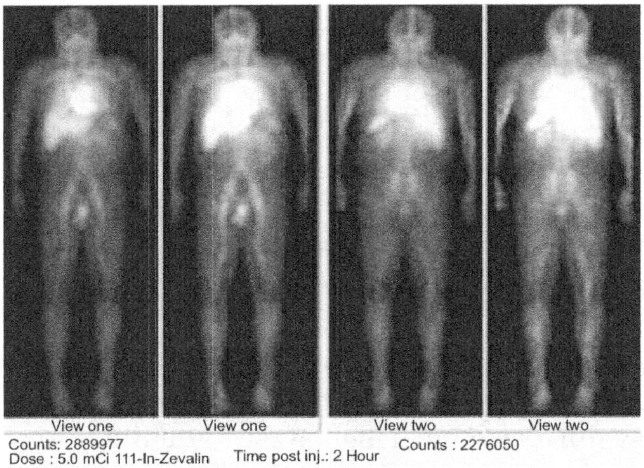

Figure 7.10 Imaging of a patient treated with Rituximab and Zevalin (www.rad.kumc.edu/nucmed/ images/Zevalin2hrwb.gif)

The whole body scan, after 2-hour post-injection, showed a normal distribution of the ^{111}In-Zevalin in the heart, lungs, liver, spleen with no **increased activity** seen in the soft tissues. On the 48-hour whole body scan, again seen the normal distribution of the ^{111}In-Zevalin in the heart, lungs, liver, spleen, but with much **less activity** seen in the soft tissues. In addition, an increased activity was observed in the left axilla, bilateral iliac regions, right greater than left, as well as within the mid-abdomen. The change in distribution from the two-hour scan to the 48-hour scan demonstrates that the patient clears the activity adequately and therefore they are a candidate for the ^{90}Y-Zevalin treatment. Patients who develop severe infusion reactions should have Rituximab, ^{111}In-Zevalin, and ^{90}Y-Zevalin infusions discontinued and receive medical treatment. ^{90}Y-Zevalin administration results in severe and prolonged cytopenias in most patients. The Zevalin therapeutic regimen should not be administered to patients with ≥25% lymphoma marrow involvement and/or impaired bone marrow reserve. Severe cutaneous and mucocutaneous reactions, some with fatal outcome, have been reported in association with the Zevalin therapeutic regimen. Patients experiencing a severe cutaneous or mucocutaneous reaction should not receive any further component of the Zevalin therapeutic regimen and should seek prompt medical evaluation.

Tositumomab (Bexxar)

It is an anti-neoplastic radioimmunotherapeutic monoclonal radiolabelled with ^{131}Iodine. It is a murine IgG2a lambda monoclonal directed against CD20 antigen found on the surface of normal and malignant B lymphocytes. On 3, February 2005, The New England Journal of Medicine reported that 59% of patients with a B-cell lymphoma were disease free for 5 years after single treatment. The monoclonal antibody used in the Bexxar therapeutic regimen called Tositumomab specifically recognizes and binds to the CD20 antigen present in the cancerous B-lymphocytes in patients with B-cell non-Hodgkin's lymphoma—the same antigen recognized Rituxan and Zevalin. Once it is labelled with ^{131}Iodine, Tositumomab is able to deliver its radiation to targeted cells. ^{131}I emits two forms of radiation: (i) Beta radiation is responsible for most of the tumour-killing effect, and (ii) gamma radiation allows gamma camera scans to be performed to evaluate the distribution and clearance of radiation from the body.

Bexxar functions by

1. apoptosis (cell-induced self killing or programmed cell death),
2. complement-dependent cytotoxicity (CDC) (antibody fixes the complement that kills the cells);
3. antibody-dependent cellular cytotoxicity (ADCC) (inducing immune effector cell killing); and

4. ionizing radiation from the radioisotope which kills cancer cell and a few adjacent normal cells by "cross fire" or "bystander" effect (Figure 7.11). ^{131}Iodine radiations are eliminated from the body mainly through the urine and by the natural decay of ^{131}Iodine.

Figure 7.11 Crossfire effect of Bexxar (www.bexxar.com/ images/ MoAright.jpg)

The US Food and Drug Administration has approved Bexxar in 2003 for the treatment of patients with relapsed or refractory follicular, non-Hodgkin's lymphoma expressing the CD20 antigen whose disease is refractory to Rituximab (not responding or a remission of less than six months) and who have relapsed following the chemotherapy. Patients who initially presented with follicular lymphoma but whose disease has subsequently "transformed" to a more aggressive type of lymphoma (e.g. diffuse large B-cell lymphoma) were also approved for therapy with Bexxar. It is not indicated for the initial treatment of patients with CD20-positive non-Hodgkin's lymphoma. Clinical trials have shown promise in a variety of other clinical settings and other B-cell malignancies, but these have not yet been approved for treatment by the FDA.

Bexxar should be administered only by the physicians and other healthcare professionals who are trained in the safe use and handling of radioactive components, i.e., either in the Nuclear Medicine Department or the Radiation Oncology Department within the hospital or clinic.

The Bexxar therapeutic regimen is delivered in two sets of intravenous infusions given 7–14 days apart: Non-radioactive Tositumomab is given before both the "dosimetric" and the "therapeutic" infusions to improve the distribution of these doses throughout the body. A trace amount of radioactive ^{131}Iodine

Tositumomab is initially given, enabling the physicians to evaluate the clearance of radiation from the patient's body using gamma camera scans. The patient then returns to the hospital for two more scans, approximately two days apart. These procedures are important because individual factors such as the tumour size, bone marrow involvement, and spleen size affect the longevity of the radiation in the body. Calculations made on the basis of these individualized radiation clearance rates allow the therapeutic dose (given 7–14 days after the dosimetric infusion) to be tailored to achieve maximum effectiveness and to minimize toxicities for each patient. The therapeutic dose contains Tositumomab labelled with ^{131}I the amount of which is specifically calculated for the patient, based on the scans performed following the dosimetric dose.

Patients with rapid clearance required a higher treatment dose than those with slow clearance to deliver the same absorbed dose (Figure 7.12).

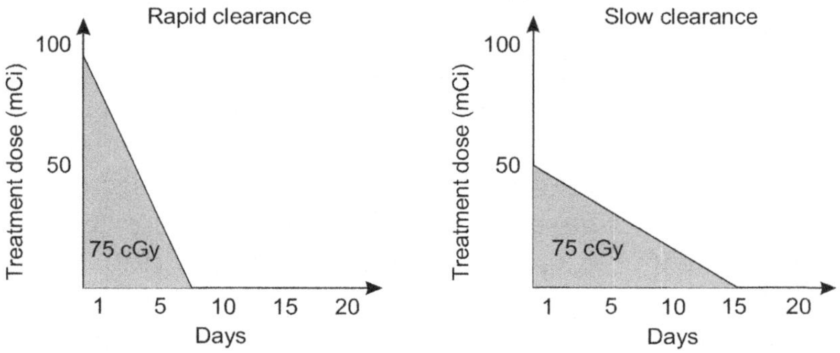

Figure 7.12 Dosimetry of Bexxar (www.bexxar.com/images/rapidclearance.gif)

Starting one day before beginning the Bexxar therapeutic regimen and continuing for two weeks after receiving the therapeutic dose, the patients take a medication containing non-radioactive iodine to protect their thyroid gland from ^{131}Iodine radiation. Once the therapeutic dose is successfully administered, the treatment is complete. In most cases, patients can receive Bexxar on an outpatient basis though this varies from state to state, dependent on local laws. Following the treatment, patients are provided with simple instructions to follow for a short period of time to minimize the radiation exposure to other people.

If these procedures are followed, the potential exposure risk to their family members is roughly equivalent to the exposure received in the course of one to two years from the normal background radiation in the environment. The amount of radiation received by close contacts of patients, who followed the instructions within the guidelines, was well deemed acceptable by the government agency, namely, the Nuclear Regulatory Commission (NRC) regulating radiation exposure.

The US Food and Drug Administration has not approved combination therapies with Bexxar; research studies suggest that treatment regimens such as CHOP chemotherapy followed by Bexxar, Fludarabine followed by Bexxar, and high-dose Bexxar followed by the transplantation of stem cell are well tolerated and effective. At the same time, though, such therapies should be conducted only on approved research protocols.

Serious hypersensitivity reactions, including some with fatal outcome, have been reported in some patients with the Bexxar therapeutic regimen. Medications for the treatment of severe hypersensitivity reactions should be available for immediate use. In the patients who develop severe hypersensitivity reactions, the infusions of the Bexxar therapeutic regimen should be discontinued and should be given proper medical attention. Severe thrombocytopenia and cytopenia have also been reported in majority of patients.

The Bexxar therapeutic regimen should not be administered to pregnant or breast-feeding mothers, to patients with >25% lymphoma marrow involvement and/or impaired bone marrow reserve, to patients have a low platelet counts (<100,000) or having a low neutrophil counts (<1500), and to those having impaired renal functions.

Cetuximab (Erbitux)

Colorectal cancer—cancer of the colon or rectum—is the third most common cancer affecting men and women. In the US, approximately 147,500 new cases were diagnosed in 2003. According to the Centers for Disease Control and Prevention (CDCP), it is the second leading cause of cancer-related death. The FDA has approved the Erbitux (Cetuximab) treatment, in 2004, to treat the patients with advanced colorectal cancer that has spread to other parts of the body. It can be given intravenously as a combination treatment with irinotecan, another drug approved to fight colorectal cancer, or alone if patients cannot tolerate irinotecan.

It is a recombinant tailored antibody targeting the epidermal growth factor receptor (EGFR) found on the surface of malignant cells of advanced colorectal cancer. It is composed of the Fv region of murine anti-EGFR antibody with human IgG1 heavy- and kappa-light-chain constant regions. It attaches itself with a high affinity to the extracellular domain of human HER-1, another EGFR, and prevents the receptor from being activated and also prevents the subsequent signal-transduction events leading to cell proliferation (Figure 7.13). Thus it has the potential to prevent cell proliferation, angiogenesis, and metastasis by promoting apoptosis.

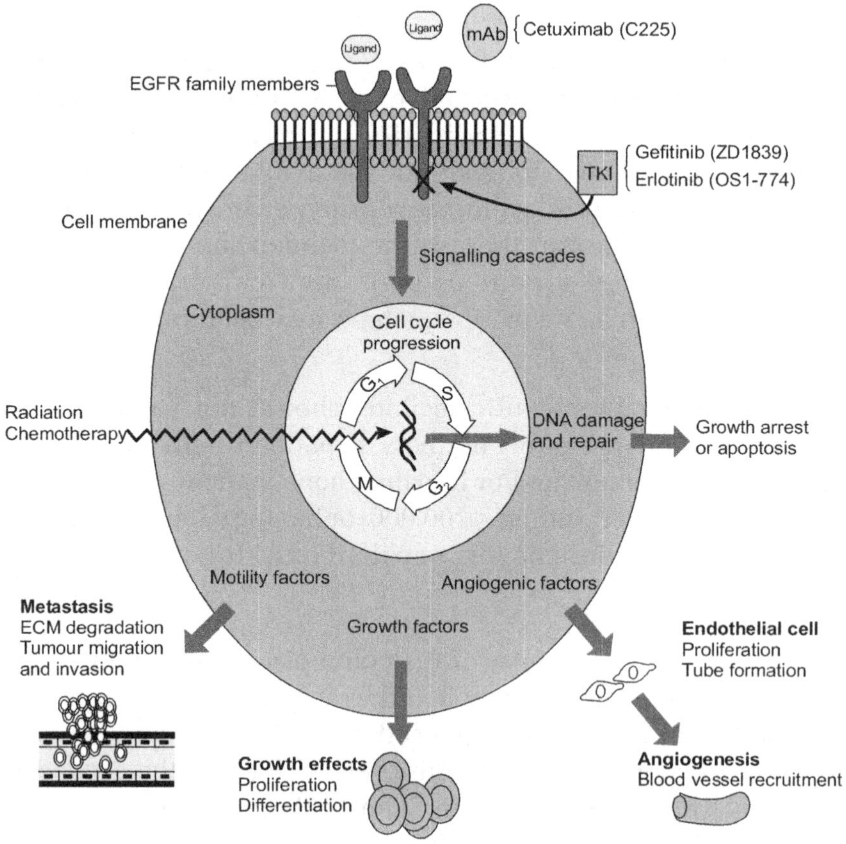

Figure 7.13 Mechanism of action of Cetuximab on cancer cell (www.humonc.wisc.edu/. ../EGFR-04-web.jpg)

It works in a different way from both the chemotherapy and hormonal therapy. Cetuximab has been evaluated alone and in combination with radiotherapy and various cytotoxic chemotherapeutic agents like Geftinib, Erlotinib, etc., in a series of phase II and phase III studies that primarily treated patients with head-and-neck or colorectal cancer. Breast cancer trials also are underway. It is also used to treat non-small cell lung cancer.

Although the FDA approval process for Cetuximab initially was slowed owing to the concerns about clinical trial design and outcome data management, the antibody was approved in February 2004 for use in combination with CPT-11 for the treatment of advanced and refractory metastatic colorectal cancer. Similar to Trastuzumab, the development of Cetuximab also included an immunohistochemical test for determining the EGFR overexpression to define the patient's eligibility to receive the antibody.

For patients with tumours that express EGFR and who no longer responded to the treatment with irinotecan alone or in combination with other chemotherapy drugs, the combination treatment of Erbitux and Irinotecan shrank the tumours in 22.9% of patients and delayed the tumour growth by approximately 4.1 months. For patients who have received Erbitux alone, the tumour response rate was 10.8% and tumour growth was delayed by 1.5 months. Recently Eribitux has acquired a new, and perhaps more ominous distinction. Erbitux is one of the most expensive cancer drugs ever made ($17,000 a month). A recent editorial in the New England Journal of Medicine calculated that adding Avastin, with a cost $4,400 a month, alone to the regimen for treating advanced colon cancer would add $1.5 billion a year in new national health costs. Erbitux, obviously, could increase national health costs by many times that figure.

Some people receiving the Erbitux injection either have a reaction to the infusion in the form of shortness of breath, dizziness, nausea, itchy feeling, or have wheezing, noisy breathing, or have a hoarse voice during the injection. Some show allergic reactions like hives, swelling of the face, lips, tongue, or throat. The other symptoms include chest pain or heavy feeling, pain spreading to the arms or shoulders, nausea, sweating, general ill feeling, urinating more or less than usual, hot and dry skin with weakness or dizziness, confusion, white patches or sores inside the mouth or on lips.

Panitumumab

Panitumumab (ABX-EGF) is used for treating EGFR-positive cancers. It targets the EGFR and blocks the ligand binding. In the preclinical mouse xenograft models, this monoclonal was found to be more potent than the mouse antibody m225, the parent of the already marketed chimeric anti-EGFR antibody, Cetuximab.

Panitumumab is a human IgG2 antibody, and because IgG2 interacts only weakly with the Fc receptors CD16 and CD32, it presumably mediates the inhibition of tumour cell growth through mechanisms other than antibody-dependent cell-mediated cytotoxicity. These could involve the blockade of ligand-induced receptor signalling and/or altered signalling directed by monoclonal binding. Initial clinical results from phase I and phase I/II studies can be compared to the clinical experience with Cetuximab. Consistent with their higher affinity, these fully human monoclonals (lacking mouse sequences) are non-immunogenic and so they require a low-dosage schedule than the chimeric antibodies.

At the selected dose level, Panitumumab is associated with diarrhoea and higher frequency of skin rashes than Cetuximab. The rashes occur most commonly on face, upper chest and back but also extended to the extremities; it is characterized by multiple follicular or pustular lesions.

Monoclonal 2F8

2F8 (Genmab, Copenhagen), the second fully human monoclonal directed against EGFR, is now in phase I/II testing for the treatment of EGFR-positive cancers. Preclinical studies of 2F8 show that, like panitumumab, it is also more potent than m225 in the mouse xenograft models. Unlike Panitumumab, this monoclonal is an IgG1 antibody and may function by eliciting antibody-dependent cell-mediated cytotoxicity in addition to blocking ligand binding and normal receptor functioning. It will be interesting to compare the clinical progress of this molecule with Panitumumab to gain some insight into the role of Fc receptor interaction on the efficacy of these drugs.

Monoclonal BL-22

It is a genetically engineered immunotoxin used to treat hairy-cell leukaemia. In hairy-cell leukaemia, the malignant cells display large amounts of CD22, an adhesion protein restricted to B lymphocytes.

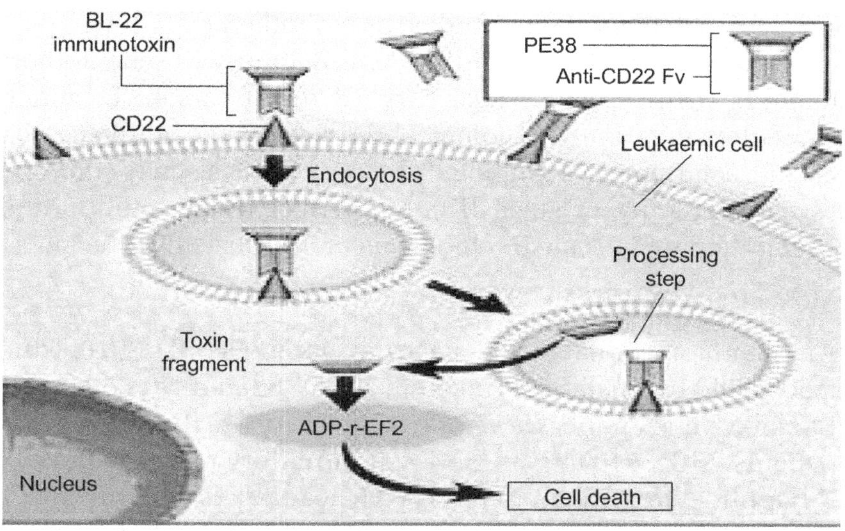

Figure 7.14 Mechanism of action of BL-22 antibody in the killing of cancer cell (deainfo.nci.nih.gov/ images/Anti-CD22.jpg)

It is conjugated with truncated pseudomonal exotoxin 38. After the conjugation with CD22, the immunotoxin becomes internalized and the toxin is released into the cytoplasm. This toxin is a glycosidase enzyme, which can modify the ribosomes and elongation factor-2. The modified ribosomes will not support the protein synthesis. One toxin molecule can modify thousands of ribosomes in one minute, and so one toxin molecule is enough to cause the apoptosis of a cancer cell (Figure 7.14).

In a preliminary study of BL-22 in patients with leukaemias and lymphomas, several patients with hairy-cell leukaemia have achieved disease remission ranging from 6 to 43 months. Of the 16 patients treated, 11 had a complete remission.

Edrecolomab

It is a murine monoclonal used as an adjuvant therapy. It targets the 17-1A antigen seen in colorectal cancer. The anti-tumour effect is mediated through ADCC and CDC. It induces anti-idiotypic network. It is undergoing investigation in large phase III trials in patients with stage III colon cancer, either as monotherapy or in combination with fluorouracil-based chemotherapy. Data from preliminary studies have shown that edrecolomab is well tolerated when used as monotherapy and adds little to chemotherapy-related side effects when used in combination.

Edrecolomab is also being studied as monotherapy following resection of stage II colon cancer, and in combination with chemotherapy in patients with resected stage II or III rectal cancer. In conclusion, edrecolomab is a novel biological therapy for the adjuvant treatment of colorectal cancer. Completed and ongoing trials may support its use as monotherapy in stage II colon cancer or in combination with chemotherapy in stage III colon cancer and stage II/III rectal cancer.

MONOCLONAL MDX-010

Another monoclonal in late-stage development, MDX-010, which was discovered by Medarex Princeton, NJ, USA, and was codeveloped by Bristol-Myers Squibb, New York, is now in phase III trials in melanoma patients. MDX-010 targets the T-cell inhibitory receptor and cytotoxic T-lymphocyte antigen 4 (CTLA-4), leading to enhanced immune responses. The experiments with hamster monoclonals directed at mouse CTLA-4 show that the resulting enhanced immune responses can mediate tumour rejection in syngenic mouse tumour models. Preclinical experiments in cynomolgus monkey models demonstrated that MDX-010 could stimulate humoral immune responses to coadministered vaccines. A published phase I/II study of MDX-010 in combination with a peptide-based vaccine in 14 stage-IV melanoma patients resulted in two complete responses and one partial response, all lasting longer than 11 months.

In addition to MDX-010, a second transgenic-mouse-derived anti-CTLA-4 monoclonal is now in clinical testing in melanoma patients.

ANGIOGENESIS INHIBITORS

Angiogenesis is the formation of new blood vessels from the pre-existing ones. It is accompanied by the neosynthesis of antigens on tumour endothelial cells. This process is a characteristic feature not only of aggressive solid tumours but also of other diseases, including rheumatoid arthritis, psoriasis and ocular disorders such as the exudative form of age-related macular degeneration and diabetic retinopathy. It is a rare phenomenon in healthy adults, occurring only locally and transiently under distinctive physiological conditions such as wound healing, inflammation and the female reproductive cycle.

In tumours, the switch to an angiogenic phenotype is known to be critical for disease progression. Unless a tumour can stimulate the formation of new blood vessels, it remains restricted to microscopic size. During angiogenesis, new antigens are formed, which are undetectable in mature vascular structures. In this light, antibodies directed against common markers of neovasculature expressed in different diseases, may open up a very general and widely applicable approach for diagnostic and therapeutic interventions. To date, however, only a few anti-angiogenesis antibodies have entered clinical investigations.

A few good-quality markers of angiogenesis are

i. integrins ($\alpha v\beta 3$, $\alpha v\beta 5$),
ii. endoglin (CD105),
iii. VEGF and VEGF-receptor complex,
iv. CD44 and ED-B of fibronectin, and
v. tenacin C.

Antibodies against these markers are angiogenesis inhibitors, which can be used either in diagnosis or in treatment. Therapeutic strategies employing targeting antibodies can be divided into two main categories. In the first approach, the antibodies are used to deliver therapeutic molecules to the vasculature. The second approach features on antibodies that may have intrinsic anti-angiogenic activity, e.g. the blocking of essential mediators of vascular proliferation.

Targeting the therapeutic molecules to tumour blood vessels often relies on the assumption that destruction and/or occlusion of tumour blood vessels may indirectly cause tumour cell death. A strong proof of this concept has been provided by Thrope and his colleagues. They used monoclonals directed against artificially induced markers on endothelial cells, to deliver the ricin toxin or a procoagulant agent to the tumour neovasculature. Complete tumour remissions were observed in a significant proportion of the mice treated. Given below are a few FDA-approved anti-angiogenic antibodies.

Bevacizumab (Avastin)

It is the first FDA-approved monoclonal which entered the US market in February 2004 and is used in combination with intravenous 5-fluorouracil-based chemotherapy as a treatment for patients with first-line or previously untreated metastatic cancer of the colon or rectum. It is designed to inhibit the angiogenesis, and thus inhibits vascular endothelial growth factor (VEGF), a protein that plays an important role in tumour angiogenesis and maintenance of existing tumour vessels. It is also designed to interfere with the blood supply to tumours (Figure 7.15) a process that is critical to tumour growth and metastasis.

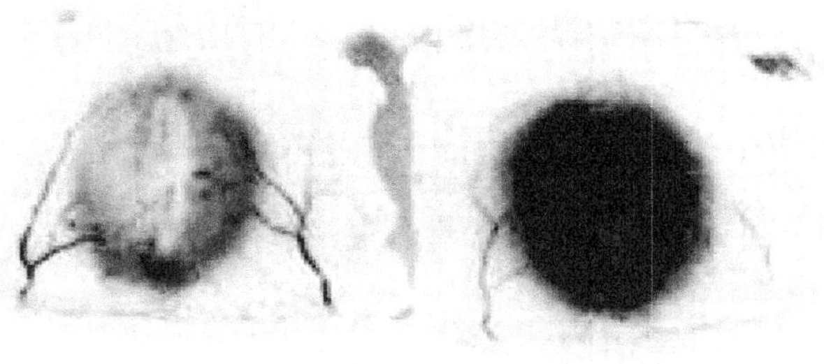

Figure 7.15 Formation of new blood vessels in a tumour, which on treatment with avastin is suppressed (*Reading from Scientific American*, 1991.)

Genentech is pursuing a broad late-stage clinical development programme with Avastin, evaluating its potential use in multiple tumour types. Phase III clinical trials in adjuvant Avastin for colorectal cancer, renal cell carcinoma, prostate cancer, and metastatic and locally advanced pancreatic cancer are being conducted. A phase III trial in first-line ovarian cancer is planned. Genentech is preparing for filing supplemental Biologics License Applications (sBLA) with the US FDA, for the use of Avastin for the treatment of first-line metastatic breast cancer, first-line metastatic non-squamous non-small cell lung cancer and second-line metastatic colorectal cancer.

Roche and Genentech in their phase II study of Avastin (Bevacizumab) plus chemotherapy in patients with metastatic colorectal cancer found a 67 per cent prolongation in progression-free survival, which was highly statistically significant. The study also showed a 29 per cent improvement in survival in patients who received Avastin plus chemotherapy compared to those receiving chemotherapy alone, but was not statistically significant. Avastin increases the response rate with chemotherapy by 9% compared to Erbitux combination therapy which is only 7% (Figure 7.16).

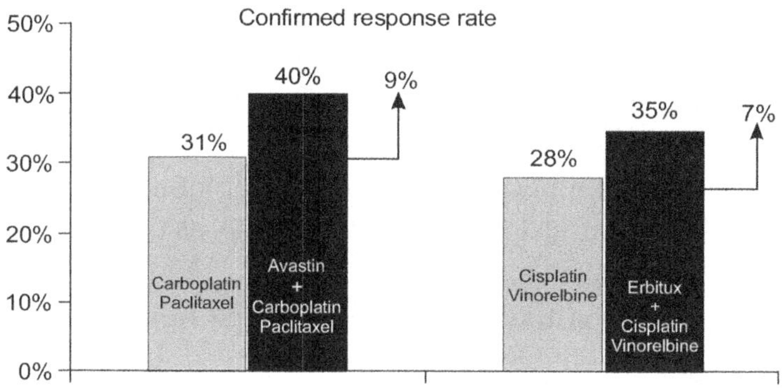

Figure 7.16 Confirmed response rate of Avastin and Erbitux combination therapy (www.agennix.com/.../ nsclc_graph02.gif)

VITAXIN

It binds to the vascular integrin ($\alpha v \beta 3$) found on the blood vessels of tumours but not on the blood vessels supplying to normal tissues. It is in Phase II clinical trials. It has shown some promising results in shrinking tumours without any harmful side effects. In the second half of 2003, MedImmune initiated two Phase II studies with Vitaxin in patients suffering from melanoma and prostate cancer. The phase II melanoma trial was a randomized, open label study, involving 110 patients with stage IV metastatic melanoma at more than 20 sites in the United States, designed to examine the safety and anti-tumour activity of Vitaxin. The phase II prostate cancer trial was also a randomized, open label, two-arm study, involving approximately 110 patients with androgen-independent prostate cancer that has metastasized to bone, designed to examine the safety and anti-tumour activity of Vitaxin in combination with chemotherapy.

OBSTACLES TO SUCCESSFUL ANTIBODY THERAPY

There has been tremendous success in the clinical applications of monoclonals in haemotologic malignancies and solid tumours. However, there are a number of obstacles to a successful monoclonal therapy. They are the following.

- Distribution of antigen present on malignant cells is highly heterogeneous.
- Density of antigen may vary from cell-to-cell.

- Some cells may express tumour antigen while others do not.
- Monoclonals are relatively fragile molecules, whose activity is affected by several factors such as the structure of monoclonal, dose and route of administration.
- If monoclonals are delivered via blood, then there may be difficulty in reaching their target site, as the tumour blood flow is not always optimal.
- High interstitial pressure within the tumour can prevent the passive monoclonals from binding.
- Sometimes tumour antigens may be released into blood, so that the monoclonals bind with free-floating antigen and not with the antigen on the tumour cells. This efficacy not only decreases the monoclonal therapy but also eliminates the possibility of treatment.
- Very rarely there is a cross reactivity, the ability of antibody to react with similar antigenic sites with normal tissue antigen. In general, target antigens that are not cross-reactive with normal tissue antigen are chosen for the effectiveness of the mouse monoclonals, hence fail to cure. To avoid this, either chimeric or humanized antibodies are used.
- Sometimes naked tumour-specific or tumour-selective monoclonals result in inefficient killing of cells. To avoid this problem, the conjugated monoclonals or immunotoxins are used.
- When the mouse monoclonal antibodies are used for treatment, since it is a foreign protein, it induces the formation of human anti-mouse antibodies (HAMA). This may diminish the activity of the therapeutic monoclonals.

TREATMENT OF LYMPHOMAS BY BONE MARROW TRANSPLANT

Patients who do not respond to chemotherapy or radiotherapy may be given bone marrow transplant, in order to supply blood-forming cells, as a last effort to save their life. In order to minimize the risk of GVH (graft versus host) disease and allograft rejection, autologous transplant is preferred. Before such transplants are performed, the patient's own bone marrow is destroyed with drug or radiation to eliminate malignant cells. Then the bone marrow is treated with Rituxan. Complement is then added to promote lysis of lymphoma cells. *In vivo* purging process of treated bone marrow removes blood stem cells of leukaemia and thus improve the effectiveness of the autologous transplant, giving patients a transplant virtually free of leukaemia, a transplant equivalent to that of healthy donor. This kind of new treatment may prolong the patient's disease-free survival, by effectively cleaning leukaemia cells or reducing them to low level. People up to 65 years of age will be able to undergo the procedure with increased safety. Appropriate monoclonals fixed to magnetic beads can also be used to get rid of cancer cells, as mentioned in the previous chapter.

MONOCLONALS IN IMMUNOSUPPRESSION

To prevent graft rejection, kidney-transplanted patients are routinely given drugs to suppress the activities of their immune systems. Nevertheless 60% of the patients who have kidney transplant, experience rejection episodes every year, that threaten the loss of the grafted organ. Graft rejection is mediated primarily by the T cells. To reduce the morbidity, mortality and cost associated with renal rejection, acute graft rejection is managed by the immunosuppressive induction therapy.

Muronomab (Orthoclone-OKT3)

To achieve a rapid augmented immunosuppression immediately following transplantation, an IgG2a monoclonal called Muronomab (Orthoclone OKT3) is given to target activation-induced antigens (OKT3 or CD3) on all T cells (Figure 7.17).

Muronomab binds to all T cells, activates earlier expression of the T cells that lead to cytokine release, and blocks the functions of T cell. It reacts with the T cells present in the most peripheral blood and with the T cells in body tissues but has not been found to react with other haemopoietic elements. A rapid and concomitant decrease in the number of circulating CD3 positive cells including CD2, CD4 or CD8, has been observed in patients studied within few minutes after its administration. After the termination of the therapy, T cells function normally within one week. In a national clinical trial, treatment with this monoclonal has saved the kidney of 90% of the patients with rejection episode compared to other immunosuppressive drugs that had only a 75% success rate. These results supported the use of monoclonals in immunosuppresson. It is also used in reversing acute cardiac and hepatic allograft in patients who are non-responsive to high doses of steroids.

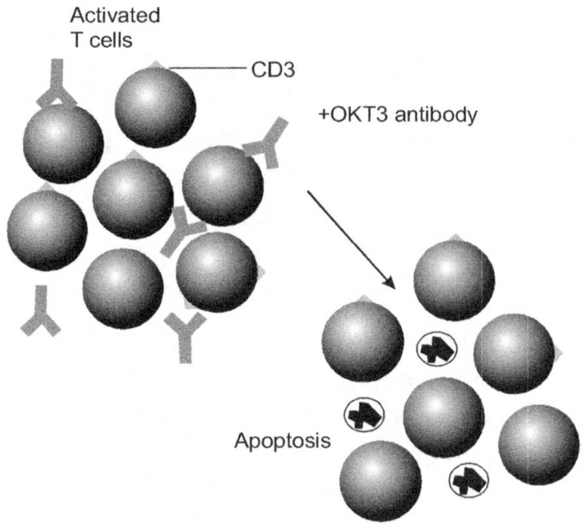

Figure 7.17 Orthoclone OKT3 binds to CD3 of the T cells of organ-transplanted patients thus suppressing the immune response (www.biology.iupui.edu/. ../images/OKT3.jpg)

Monoclonal antibodies are not only effective than drugs in kidney rejection, but also less prone to causing side effects. The monoclonals suppress only the T cells, whereas the drugs like cyclosporine and corticosteroids suppress all aspects of immune function, thereby leaving the patients more prone to infection. In addition, drugs may damage the kidney permanently. Orthoclone OKT3 can cause reversible, easily controllable and transient side effects.

Only the physicians experienced in immunotherapy and management of solid organ transplant patients should use Muronomab, since patients occasionally face serious life-threatening or lethal reactions. Release of numerous cytokines/lymphokines are responsible for many acute clinical manifestations like cytokine released syndrome a common complication occurring with the use of anti-T-cell antibody for serious events in central nervous system like aseptic meningitis, for opportunistic infections/viral induced lymphoproliferation, for increased risk of neoplasia due to increased cell-mediated immune response, and for thrombosis, hypersensitivity and HAMA reactions.

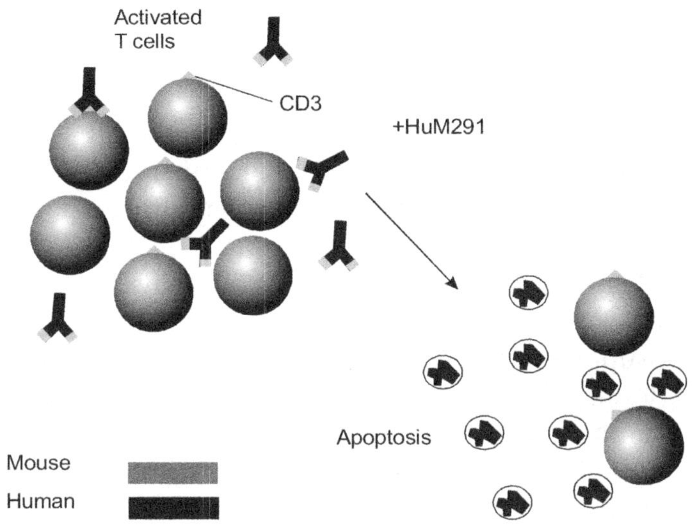

Figure 7.18 Humanizing the OKT3 antibody and stripping 90% of the characteristics of mouse increase the destruction of activated T cells and the half-life (www.biology.iupui.edu/. ../images/ OKT3.jpg)

Seeking to improve on the OKT3, a mouse antibody called BC3 was produced. This drug also suppresses GVHD. Although BC3 produced milder side effects than OKT3, its overall effectiveness is still not confirmed.

Humanizing the OKT3 antibody, stripping of 90% of mouse characteristics and further engineering the tail helped to prevent it from non-specifically activating T cells. The antibody named HuM 291 (Figure 7.18), effectively blocks the T-cell activity. This antibody not only inactivates the naive T cells but also efficiently suppresses the activated T cells. It also selectively kills the activated T cells. An added advantage of this humanized antibody is its increased half-life, and hence it requires a less frequent dosage. Also this antibody drug is in trials for treating the autoimmune diseases like psoriasis and multiple sclerosis.

Researches on obtaining new immunosuppressive monoclonals with different target specificity, effective pathophysiological processes, and minimal immunogenic and infusion-related problems, and being safer to use and more efficacious even in a short course of administration, have led to the discovery of Basiliximab and Daclizumab.

Basiliximab (Simulect)

Basiliximab is a chimeric monoclonal that retains the murine variable region. It is a recombinant monoclonal against a trimeric cell surface receptor IL-2a expressed only on the T lymphocytes activated by the interaction with a foreign antigen or with free IL-2 and not on resting T cells. The IL-2a subunit (CD25, T-cell activation antigen, Tac) is particularly important from the standpoint of selective immunosuppressant action because it is the only receptor component specific for IL-2. The murine monoclonals (anti-Tac) initially developed were immunogenic to human leading to diminished efficacy and to low tolerability with adverse and hypersensitive reactions and had a relatively short half-life thereby requiring repeated dosing. Modification of this rodent-anti-Tac by recombinant technology has led to the development of marketing Basiliximab.

It inhibits the IL-2-mediated activation and proliferation of T cells, the critical step in the cascade of cellular immune responses of allograft rejection. It has a long half-life of approximately 7–12 days and saturates IL-2 for 59 days. Less adverse effects were seen when used in standard immunotherapy because it is highly humanized. Several clinical studies have shown that the administration of this monoclonal as an induction agent significantly reduced the incidence of acute rejection with minimal side effects, even in high-risk patients. During the 1st year of post-transplant period, the addition of Basiliximab to the standard immunosuppressive regimen resulted in an ~30% reduction in the incidence of acute allograft rejection.

Daclizumab (Zenapax)

It is a humanized monoclonal against IL-2 receptor (anti-Tac) (Figure 7.19), which retains only the hypervariable region of murine antibody, developed to reduce the immunogenicity. *In vitro* experiments demonstrated that daclizumab has approximately one-tenth the affinity of Basiliximab for IL-2a protein, with a saturated concentration of 5–10 mg/ml and 0.1–0.4 mg/ml respectively. These differences in their target affinity influence the dosing schedules of the two monoclonals.

Its half-life is 20 days and it saturates IL-2 receptor for up to 120 days.

During the first six months post-transplant period, the addition of Daclizumab to a standard immunosuppressive regimen resulted in 50% reduction in the incidence of acute allograft rejection, compared to 30% reduction observed in Basiliximab. The most commonly reported side effects are urinary tract infection and CMV. No incremental increase in adverse reactions has been noted when these new monoclonals were used with other common immunosuppressants. Daclizumab stimulated the anti-idiotypic antibodies in 8.4% of cases while Basiliximab stimulated such antibodies in 0.4% of patients.

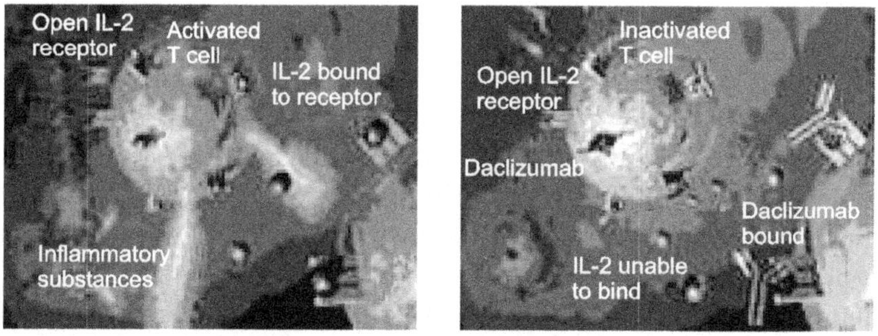

Figure 7.19 Immunosuppression of daclizumab (a) IL-2 of a normal T cell and its receptor (b) Inactivation of T cell by binding with the IL-2 receptor. (www.artemis-creative.com/image/port/i_pdl-dac...)

Daclizumab is also being investigated for treating inflammatory and autoimmune diseases, including asthma and multiple sclerosis.

There has been noticeable evolution in the selection of immunosuppressive agents in solid organ transplantation over the past 10 years (Figure 7.20). An ongoing shift is seen in the types of antibody preparation being utilized as induction therapy, from the Muronomab-CD3 and horse ATGAM to the rabbit antithymocyte globulin and to the monoclonals—anti-IL-2 receptor antagonists—Daclizumab and Basiliximab.

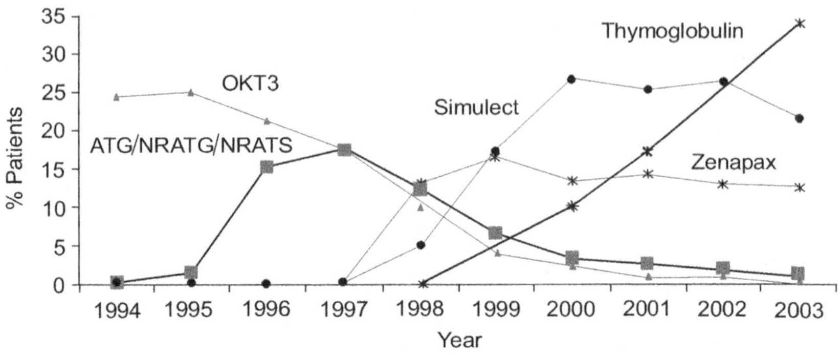

Figure 7.20 Graph illustrating the trend in the use of immunosuppressive drugs in kidney transplantation

MONOCLONALS AGAINST CEREBRAL OR CORONARY THROMBUS

The majority of the natural deaths in the world are due to the consequences of blockage of either a cerebral or a coronary artery by a blood clot or thrombus. The thrombus consists of a network of fibrin—a blood-clotting agent formed in response to a defect in the wall of a blood vessel. Under natural conditions, plasmin (a serine protease, produced by the activation of plasminogen by plasminogen activator) degrades the fibrin in the blood clot and causes the clot to dissolve (Figure 6.22). However, in many instances, arterial blockage is caused when this biological system does not remove the blood clots efficiently. Thus plasminogen activators could be used as therapeutic agents to induce higher levels of plasmin which in turn would efficiently remove the arterial thrombi, thereby reducing their impact on arteries of brain and heart.

Plasmin can also degrade fibrinogen which is a precursor of fibrin, and so this therapy can lead to an excessive internal bleeding as a result of depletion of the fibrinogen levels. Consequently, thrombolytic agents that can degrade only the fibrin present in the blood clots have been developed. Researchers reasoned that an antibody covalently coupled to a plasminogen activator can be used both to target the fibrin, and subsequent to binding, causes an increase in the localized levels of plasmin by converting plasminogen to plasmin. Therefore, tissue-specific plasminogen activator coupled to an antifibrin monoclonal, can bind to the blood clot and also can cause the lysis of the blood clot (Figure 6.22).

Abciximab (Reopro)

Heart attack occurs when a blood vessel in the heart is blocked by a blood clot. Blood clots can sometimes be formed during a percutaneous coronary intervention (PCI), a procedure to open the blocked arteries of the heart. Abciximab is used to lessen the chance of heart attack in people who need this PCI procedure. In other words, it is helpful in preventing the reclogging of the coronary arteries in patients who have undergone angioplasty. It is a chimeric monoclonal directed against the platelet glycoprotein IIb/IIIa receptor (Figure 7.21). It inhibits the clumping of platelets by binding the receptors on their surface that are normally linked by fibrinogen, and by preventing the fibrinogen and VW factor from initiating the platelet aggregation. It also inhibits fibronectin receptor (Vα3β) which mediates both the coagulation of platelets and the proliferation of endothelial and vascular smooth muscle cells. Abciximab is used along with the aspirin and heparin, which are other medicines used to prevent the blood from clotting.

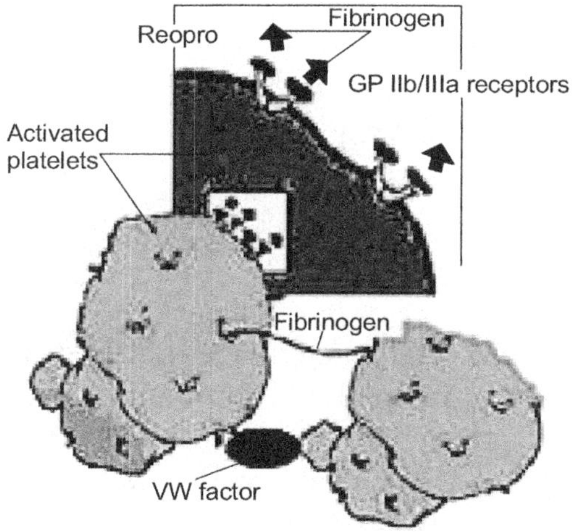

Figure 7.21 ReoPro binds with GP IIb/IIIa receptor and inhibits clumping of platelets (www.nscardiology.com/plt5.gif)

It has been approved as an adjunctive therapy to prevent cardiac ischemic complications, as well as in unstable angina patients not responding to conventional medical therapy when PCI is planned within 24 hours. ReoPro, made by Centocor in Malvern, Pa., had a last-year sales of $418 million. Previously, ReoPro was indicated solely for use in angioplasty patients at high risk for complications and was administered as a bolus—10–60 minutes before the PCI procedure, followed by a 12-hour continuous infusion. The expanded labelling allows the physicians to use ReoPro in a broader population of patients undergoing balloon angioplasty, atherectomy, or stent placement. In addition, in patients with refractory unstable angina, ReoPro can now be administered 18 to 24 hours before PCI.

Several companies are also pursuing monoclonals against CD18, a protein on T lymphocytes, which underlies inflammation as well as tissue damage resulting from a heart attack.

MONOCLONALS AGAINST AUTOIMMUNE DISEASES

Autoimmune diseases are the diseases that affect the specific physiological function of any organ or whole body due to complexes deposited by self (auto) antigens and their (auto) antibodies. They cause tissue damage and inflammation. The ultimate goal of immunotherapy for autoimmune diseases is to find a specific intervention to restore the tolerance to the relevant autoantigen. Two therapeutic approaches under investigation are the following.

i. Identifying and blocking the TCR or Ig on lymphocytes to which the autoantigen bind.
ii. Identifying and blocking the MHC class I and class II molecules responsible for presenting peptides from autoantigens.

Infliximab (Remicade)

This is a very successful monoclonal antibody drug now on the market. The body uses this monoclonal to enhance the cellular arm of the immune system and also to cure the inflammatory diseases. According to company reports, Remicade, which is on pharmacy shelves for Crohn's disease (an inflammatory bowel disease) and rheumatoid arthritis, made $370 million last year for its developer, Centocor. Therapies that wipe out TNF have a potential $2 billion annual market, according to Carol Werther, Managing Director of Equity Research at the Investment Bank Adams, Harkness and Hill.

In recent years, a large number of proteins, termed cytokines, have been identified as intercellular messengers in inflammation, immune response, and tissue repair and remodelling. Cytokines have been classified as interleukins, interferons, colony stimulating factors, growth factors and tumour necrotic factors (TNF). Some cytokines show strong anti-inflammatory activity and are utilized to reduce tissue injury in many inflammatory diseases.

The two kinds of TNF that exist are TNF-α and TNF-β. TNF-α is responsible for the induction of pro-inflammatory cytokines like IL-1 and IL-6 and for increasing the endothelial permeability, thereby enhancing migration of the leucocytes, expression of adhesion molecules by the endothelial cells and leucocytes, activation of neutrophils and eosinophil cells, proliferation of the fibroblast synthesis of prostaglandins, and the induction of acute-phase and other liver proteins.

TNF-α is produced by monocytes. Initially, it is expressed as a 26-kDa membrane protein, which is subsequently cleaved by an enzyme to a 17-kDa soluble monomer. Association of these monomers into a homotrimer yields the active TNF-α. Its receptor TNF-R1 is proposed to mediate the TNF-α's action related to inflammation. The monoclonal against TNF-α binds with both the free TNF-α trimers (Figure 7.22a) and membrane-bound cytokines, and neutralizes the pathophysiological action of TNF-α. These antibodies induce temporary remission of rheumatoid arthritis, colitis, synuvitis and polyarthritis. The presence of anti-TNF-α is also reported to suppress the production and secretion of TNF-α. One such antibody Infliximab marketed as Remicade is approved for the treatment of moderate to severe active Crohn's disease (Figure 7.22b). In combination with methotrexate, it is used for the treatment of rheumatoid arthritis. It is also used for the treatment of ulcerative colitis and Behcet's syndrome.

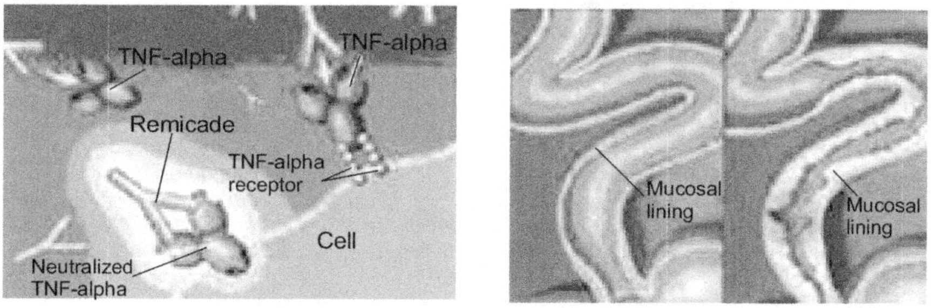

Figure 7.22 (a) Remicade inactivation TNF-α (b) Mucosal lining in normal and Crohn's patients (www.remicade.com/.../how_remicade_ works.jpg)

It is the first monoclonal to register for use in RA (rheumatoid arthritis), and now Crohn's disease, lupus erythymatosus, spondyloarthropathies, and juvenile arthritis are also being treated. Infliximab is a chimeric (25% murine and 75% human) monoclonal of IgG1 subclass. It is genetically engineered using the variable region of the murine monoclonal. To avoid the HAMA effect, it is humanized to 100% (Adalimumab), and it is also adopted as a fusion protein (Etanercept) as shown in Figure 7.23.

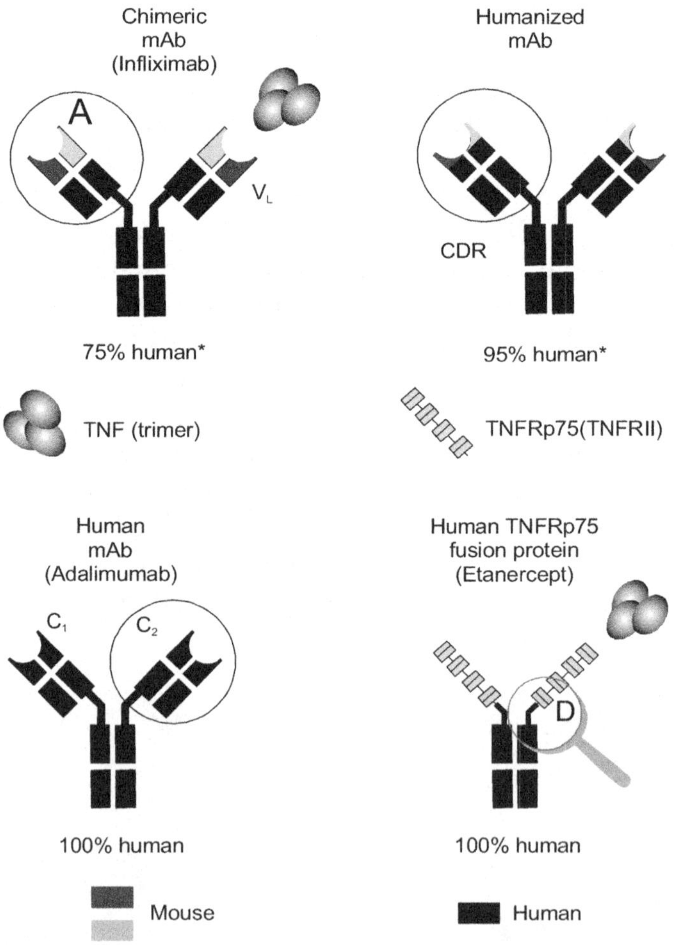

Figure 7.23 Humanization of Infliximab (www.iir.suite.dk/IIR/70TNF/images/70antiTNFα.gif)

The medication is given as an infusion, the dose determined by the weight and given at baseline, 2 weeks, 6 weeks, then bimonthly. It is used with methotrexate to prevent the antibody production to the drug itself (which is a chimeric antibody). The drug response is dramatic with a reduction in inflammatory markers, and in swollen and tender joint counts. The drug however requires the ongoing use and withdrawal results in the resurgence of disease.

Infliximab treatment of Crohn's disease patients, who had not responded adequately to conventional therapies, elicited a response rate in 80% of patients within 4 weeks of drug administration. Nearly 50% of the Infliximab-treated patients demonstrated the evidence of clinical remissions after 4 weeks of drug treatment. Infliximab is also indicated as treatment for patients demonstrated with fistulizing Crohn's disease (FCD). This new drug is the first product documented to reduce the number of open fistulas in controlled clinical trials. Clinical studies of Infliximab showed impressive and rapid effects in established rheumatoid arthritis. The response to Infliximab is dose dependent, with a single dose of 10–20 mg/kg able to generate clinically significant responses that persist for atleast one month in 80% of patients. Retreatment is effective.

EFALIZUMAB (RAPTIVA)

Of the several monoclonal antibodies under investigation, a humanized monoclonal called Efalizumab (Xoma and Genentech in the USA, Serono in Europe), was authorized in over-18s in the USA for moderate-to-severe plaque psoriasis.

Psoriasis is a disease of the skin in which the cells of the skin reproduce more rapidly than normal. This increased reproduction is thought to be stimulated by the activation of the immune system, in particular, the T lymphocytes (Figure 7.24). As a result, the skin shows a scaly appearance. In order to become activated, the lymphocytes must adhere (attach) to other cells through their receptors on the surface. Efalizumab blocks one of these receptors called leucocyte functional antigen-1 or LFA-1 or CD11a on the lymphocyte. By blocking the adhesion, Efalizumab selectively and reversibly blocks the activation, reactivation and trafficking of T cells that lead to the development of psoriasis. The trials showed that it reduced the thickness, scaling and redness of psoriatic plaques.

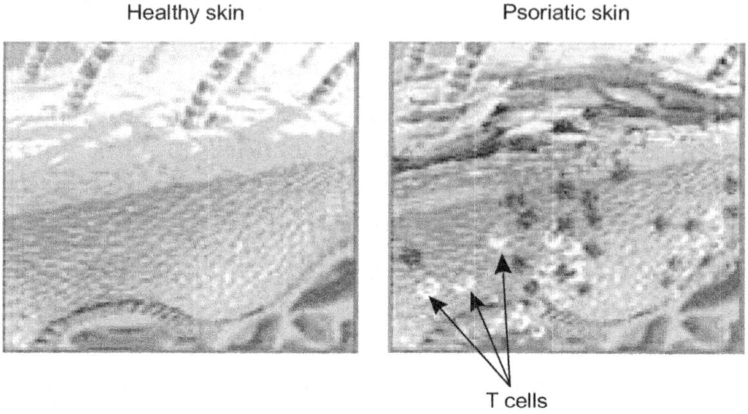

Figure 7.24 Normal healthy skin and psoriatic skin where T cells become activated (www. hybridmedicalanimation.com/images/illust_...)

The FDA approved (October 2003) Raptiva as one-dose, once-a-week self-injection, for the treatment of chronic moderate-to-severe plaque psoriasis. Subcutaneously (under the skin) injected, Raptiva treats psoriasis without multiple oral medications or frequent topical applications, and similar to taking insulin for diabetes, taking Raptiva at home, can become a natural part of a psoriasis patient's weekly routine. This immunosuppressive recombinant humanized monoclonal of IgG1 kappa isotype prevents the T cells from binding with antigen-processing cells. By inhibiting this interaction, it blocks the underlying physiological mechanism that causes the skin to develop inflammation and lesions.

It is approved for the treatment of chronic, moderate-to-severe plaque psoriasis in adults (18 years or older) who are candidates for systemic therapy or phototherapy. An estimated 4.5 million Americans have psoriasis and of these, 1.5 million are considered to have moderate-to-severe disease. In clinical studies, Raptiva demonstrated a rapid onset of action in some patients by four weeks and in the reduction of symptoms associated with psoriasis.

Raptiva was approved on the basis of efficacy and safety seen in 4 pivotal trials, which showed a sustained improvement in psoriasis symptoms over 6 months. Genentech presented the final results from a long-term study that showed sustained improvement in psoriasis symptoms throughout three years of continuous treatment, at the meeting of the American Association of Dermatology Academy, 2005.

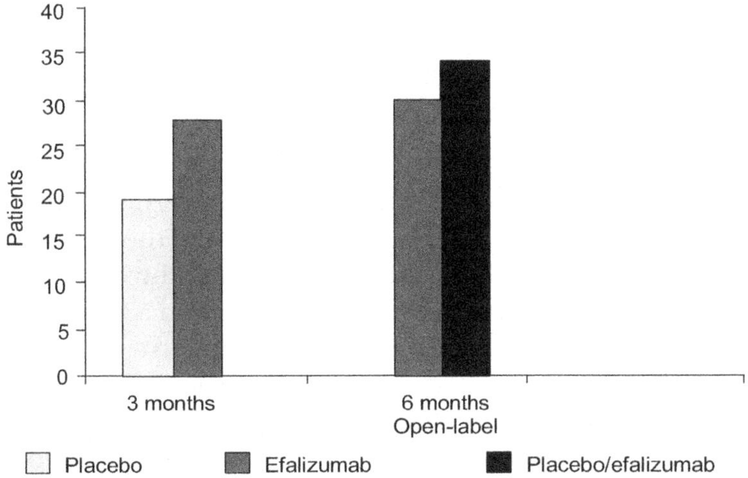

Figure 7.25 Response of Efalizumab in psoriatic arthritis patients given at 3 months and 6 months (www.hybridmedicalanimation.com/images/illust_...)

In the psoriatic arthritis patients, Efalizumab blocks the interaction of LFA1-ICAM1 and shows a good response at 3 and 6 months (Figure 7.25).

Two other monoclonals are designed, like Etanercept, to target the TNF-α. One is Infliximab (Schering-Plough), a humanized monoclonal, which is in phase II clinical trials. The other, Adalimumab (Humira-Abbott/CAT), is a fully human monoclonal discovered in collaboration between the British Biotechnology Company Cambridge Antibody Technology and Abbott. Abbott is responsible for its clinical development and already has an authorization for its use in severe rheumatoid arthritis. Adalimumab is now in advanced trials for psoriasis and psoriatic arthritis. It binds to TNF-α and specifically neutralizes TNF-α rather than blocking its receptor.

The importance of TNF-α in psoriasis is unquestioned, but other companies have developed products that target different parts of the disease cycle. For example, there are evidences that several of the interleukins contribute to skin damage in both psoriasis and eczema. A fully human monoclonal, ABX-IL8 (Abgenix), neutralizes IL-8, which can be raised by 150-fold in psoriatic tissue. Another fully human monoclonal, HuMax-IL15 (Genmab and Amgen), neutralizes IL-15. Both of them have reached phase II clinical trials and have demonstrated clinical activity. Yet another chemical messenger called interferon-gamma is the target of the humanized monoclonal, Fontolisumab (Protein Design Laboratories). This monoclonal has reached only the phase I trials, but could provide evidences for a key role of interferon gamma in psoriasis.

Another group of potential medicines act like Alefacept, in that they are designed to block various cell receptors. Among them, HuMax-CD4 (Genmab) is a human monoclonal against the CD4 receptor. In a Phase II trial, almost one-third of patients showed a 25 per cent reduction in the area and severity of disease and 53 per cent showed a sustained benefit for 12 weeks after treatment had ended. A subsequent phase II trial is expected to start shortly. Another humanized compound, Siplizumab (MedImmune), binds to a different receptor called the CD2 receptor. It has completed phase I/II studies but has been put on hold while the company pursues an alternative anti-psoriasis compound.

Eculizumab (Alexion), the monoclonal antibody presently in development for psoriasis, has a somewhat different mechanism of action. It is designed to stop the activation of a series of blood proteins called the complement cascade. The activated complement is shown to be present in the inflamed tissue and in the skin involved in psoriasis and has the potential to cause inflammation and skin damage. This humanized monoclonal inhibits the step that converts the complement C5 into inflammatory fragments, C5a and C5b–9, and is expected to dampen down active psoriasis. The compound is still at the early stage of clinical development.

Many other monoclonals are now in clinical trials. For example, Genentech is in the late stages of testing, its monoclonal Xanelim against CD11a. This protein exists on the surfaces of T lymphocytes and helps them to infiltrate the skin and causes the inflammation, characteristic of psoriasis. In a study of nearly 600 psoriasis sufferers that was reported at the American Academy of Dermatology Conference in July, researchers found that 57 per cent of patients on the highest dose of the drug experienced at least a 50 per cent decrease in the severity of their disease.

Enbrel (Etanercept)

Enbrel, a monoclonal against tumour necrosis factor (TNF), is used for the treatment of rheumatoid arthritis, polyarticular-course juvenile rheumatoid arthritis, ankylosing spondylitis, psoriatic arthritis and psoriasis. In these patients, the level of TNF is high. The drug reduces the amount of TNF to normal levels. But in doing so, it reduces the ability of immune system to fight against infections. For the patients of rheumatoid arthritis, it can be given in combination with methotrexate (MTX). In the moderate-to-severe active condition, it reduces the symptoms, inhibits the progression of structural damages and improves the physical functions of patients. In the medical studies, 2 out of 3 adults, who used it, were found to be benefited within 3 months.

Enbrel has been studied in 1442 patients with rheumatoid arthritis, followed for up to 80 months, in 169 patients with psoriatic arthritis for up to 24 months, in 222 patients with ankylosing spondylitis for up to 10 months, and 1261 patients with plaque psoriasis for up to 15 months. In the controlled trials, the proportion of Enbrel-treated patients who discontinued the treatment due to adverse events was approximately 4% in the indications studied. The vast majority of these patients were treated with 25 mg SC (subcutaneous) twice weekly. The doses of this drug studied for plaque psoriasis were 25 mg SC once a week, 25 mg SC twice a week, and 50 mg SC twice a week (Figure 7.26).

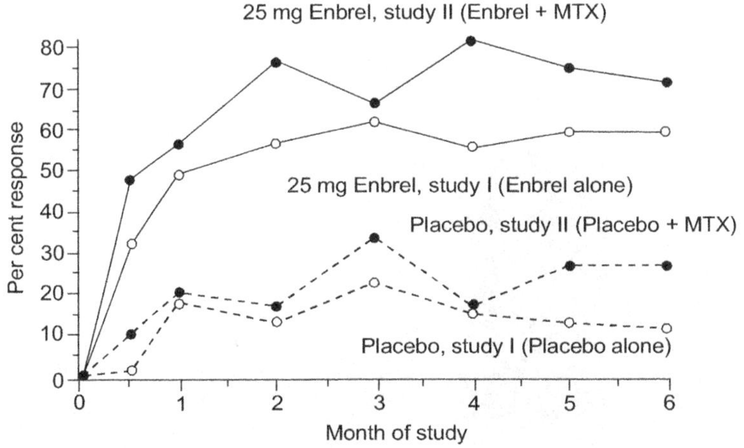

Figure 7.26 Response of Enbrel alone in study I, and in study II enbrel given along with MTX (Kuby, J., 1997)

Campath-1H

Campath-1H is an experimental drug treatment for multiple sclerosis, organ transplant rejection and several types of leukaemia. In one small trial of 27 people with secondary progressive multiple sclerosis, the drug was found to virtually eliminate the formation of new lesions and the inflammation associated with the disease for at least eighteen months. These results were demonstrated by MRI scan.

Campath-1H works by destroying the body's T cells that are believed to be responsible for initiating the destructive process seen in multiple sclerosis (MS). MS is a chronic and often a disabling disease of the brain and spinal cord. Clinical trials with 60 multiple sclerosis patients have been treated at Addenbrook's hospital at Cambridge, with Campath-1H. After five days of treatment, the occurrence of new brain lesions was greatly reduced for many years (Figure 7.27).

Treatment with Campath-1H alone gave good results in rheumatoid arthritis. Most patients entered the remission or stable disease, and the duration of the remission varied from a few weeks to several years. Repeated treatment with Campath-1H is possible and effective, but some patients eventually made an anti-idiotype response against the humanized monoclonal antibody. However, the depletion was so effective that T cells do not return to normal levels for a very long time.

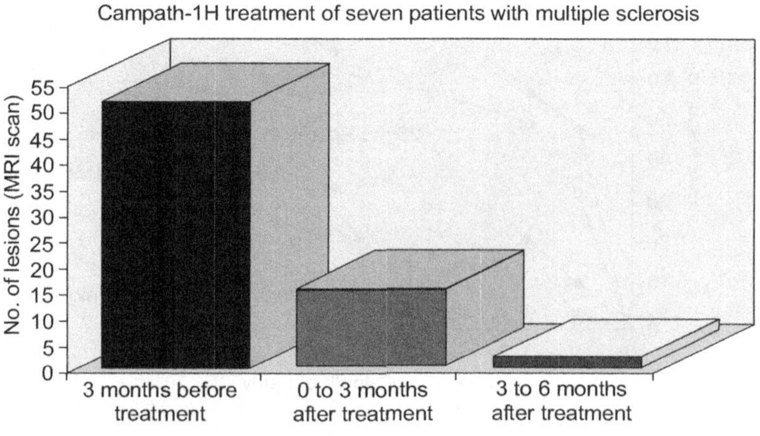

Figure 7.27 Diagram showing the Campath-1H treatment for reducing the number of brain lesions (users.path.ox.ac.uk/~scobbold/tig/Swdsp2.gif HMS Beagle (BioMedNet))

Anti-CD4 Antibody

Monoclonal antibodies have been used successfully to treat autoimmune diseases in animal models of systemic lupus erythematosus, rheumatoid arthritis and myasthenia gravis. One example is, high percentage of F-1 mice given weekly injections of high doses of monoclonals specific for CD4 membrane molecule recovered from nude mice, in which the treatment with anti-CD4 monoclonal led to the disappearance of the lymphocytic infiltration and diabetic symptoms. It also seems to markedly reduce the frequency of glomerulonephritis, vasculitis and auto-antibody formation in many systemic autoimmune diseases (Figure 7.28).

Hu5A8, a humanized monoclonal antibody against CD4, was previously shown to inhibit the replication of HIV and SIV *in vitro* and was safely administered to rhesus monkeys without depleting the $CD4^+$ T cells. Treatment with this anti-CD4 monoclonal antibody also suppresses the *in vivo* replication of HIV. Anti-CD4 antibodies, which cause $CD4^+$ T-cell depletion, have been shown to increase the susceptibility to infections in mice. Thus, the development of anti-CD4 antibodies for clinical use raises the potential concerns about suppression of host defence mechanisms against pathogens and tumours. The anti-human CD4 antibody, Keliximab, which binds only with the CD4 of humans and chimpanzee, has been evaluated in host defence models using murine CD4 knockout-human CD4 transgenic (HuCD4/Tg) mice.

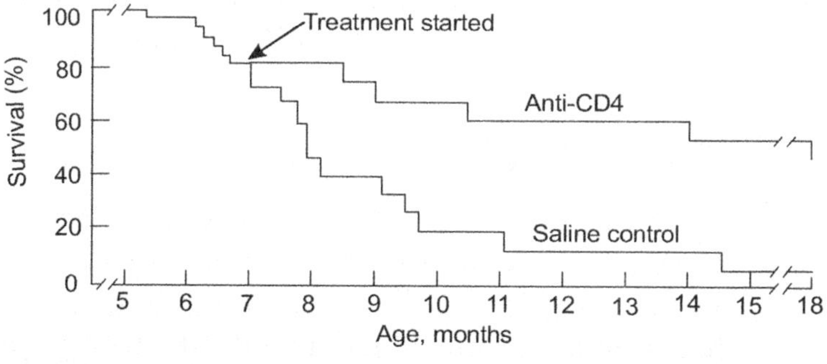

Figure 7.28 Effect of CD4 antibody increases the % of survival compared to the control mice with saline

The most recent transgenic-mouse-derived human monoclonal to enter into the late-stage clinical development is Zanolimumab

(HuMax-CD4; Genmab, Copenhagen), which binds to the T cell-differentiation antigen (CD4). A phase III clinical trial in patients with CD4 cutaneous T-cell lymphomas has been announced; however, the published clinical study in psoriasis patients, was directed towards inflammatory and autoimmune indications. The observed dose-dependent decrease in circulating CD4 cells in this study may translate to efficacy for treatment of T-cell lymphomas, where the drug is currently being developed.

NATALIZUMAB (TYSABRI)

Natalizumab, the new product, is a bioengineered monoclonal antibody from part of a mouse antibody to closely resemble a human antibody. It is being marketed under the trade name Tysabri by the company Biogen Idec' and Elan. The FDA licensed this new biologic approach to treat patients with relapsing forms of multiple sclerosis to reduce the frequency of symptom flare-ups or exacerbations of the disease. Following the reports of progressive multifocal leucoencephalopathy (PML), it was later withdrawn from the market owing to safety concerns.

In February 2005, the FDA removed the hold on clinical trial dosing of Tysabri in multiple sclerosis in the United States. The companies resubmitted this antibody to the FDA in September 2005, and it was subsequently granted priority review. Tysabri is scheduled to be reviewed by an FDA advisory committee in March 2006 and may reach the market again. This drug brought a great hope and anticipation to the public and physicians, and the PML cases were a serious setback that re-enforced how difficult it can be both to fully understand a drug's effects and to isolate the causes of conditions that affect seriously ill patients.

OTHER MONOCLONALS

Monoclonals to IL-2Rα subunit—expressed at higher level on autoimmune T cells—when injected with MBP (myelin basic protein) along with anti-TAC, block the autoreactive T cells in rats with experimental autoimmune encephalomyelitis (EAE). 6 out of the 9 rats treated had no symptoms. Injection of PL/J mice with monoclonals specific for $V_\beta 8.2$ T-cell receptors, prevented the induction of EAE by MBP in adjuvant (Figure 7.29). Even more promising was the finding that anti-$V_\beta 8.2$ monoclonal could also reverse the symptoms of autoimmunity in the mice manifesting induced-EAE. Simultaneous blockade of interaction of B7 of APC and CD28 of T cells and co-stimulatory interaction of CD40 of APC with CD40L of T cells, inhibit the synthesis of antibodies which in turn provide a long-lasting suppression of autoimmune diseases.

At least six additional transgenic-mouse-derived monoclonals are in phase II clinical trials, of which two are directed against targets, CD20 and tumour necrosis factor α(TNFα), for which non-transgenic-mouse-derived antibodies are already on the market.

Many monoclonals are in clinical trials or tried in animal models. Anti-α 4 integrins and altered peptide ligands of myelin basic protein (MBP 83-99) showed initial promising results. The use of anti-CD20 monclonal for the *in vivo* depletion of B cells and early trials of autologous peripheral stem cell transplant represent additional immunomodulatory treatment modalities for systemic lupus erythematosus patients.

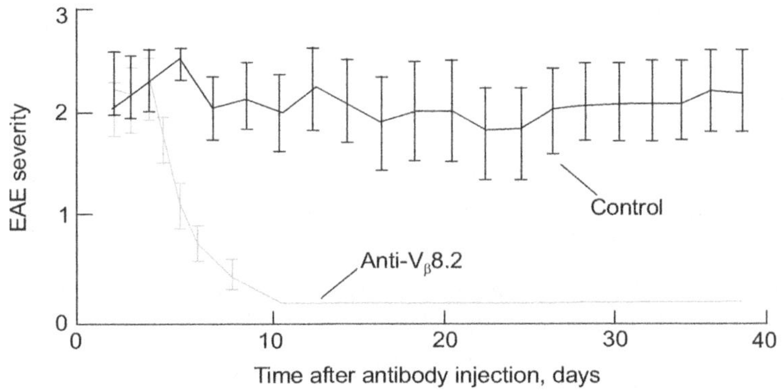

Figure 7.29 Anti-V_β 8.2 decreases EAE severity in mice. (Kuby, J., 1997)

Immunotherapy based on the identification of particular MHC molecules that drive an autoimmune response, is much more likely to be effective, but such therapy would also inhibit some protective immune responses. Injecting the mice with monoclonals to class II MHC molecules before injecting MBR, blocks the development of EAE. If instead the antibody was given after the injection of MBP, development of EAE was delayed, but the induction of symptoms of asthma such as coughing, sneezing, watery eyes, and shortness of breath was not prevented. In non-human primates, monoclonals to HLA-DR and HLA-DQ have been shown to reverse the EAE. Researches at Stanford University School of Medicine in Palo Alto, California, have found that such antibodies in mice can suppress the development of conditions similar to myasthenia gravis, multiple sclerosis and systemic lupus erythematosus of humans.

Use of monoclonals for the treatment of human autoimmune diseases presents clearly the existing possibilities. Better understanding of autoimmune conditions and advances in the production of humanized monoclonals promises a better immunotherapy in the near future.

MONOCLONALS AGAINST ALLERGIES

Until recently, the management of allergy has been limited to three approaches: **avoidance** of the allergen, pharmacologic interference with the consequences of mediator release, and **desensitization** by repeated introduction of subclinical doses of the allergen. Each of these approaches can be partially effective, but morbidity and, in rare cases, death from the allergic disorders continue to be a major public health problem.

Omalizumab (Xolair)

IgE antibodies are the key players in allergic reactions. These antibodies prompt the other cells (mast and basophil cells, among others) to begin a complex chain reaction that culminates in allergy and asthma, presenting with symptoms such as coughing, sneezing, watery eyes, and shortness of breath. The FDA approved Xolair in June 2003, in the US. It is designed to target the IgE antibodies, the key underlying the cause of the symptoms of allergy-related asthma. Xolair is a recombinant DNA-derived humanized IgG1.

It blocks the high-affinity Fc receptor of IgE antibodies, stops the free serum IgE from attaching to mast cells and other immune cells and prevents the IgE-mediated inflammatory changes (Figure 7.30). Omalizumab treatment downregulates the expression of a dendritic cell, FcεRI. The combined effect of specific immunotherapy (SIT) and Omalizumab prevented an increase in nasal eosinophilic cationic proteins and decreased the tryptase levels in nasal secretions. Anti-IgE therapy in children with allergic rhinitis reduces the release of leukotriene from the peripheral leucocytes stimulated with the allergen.

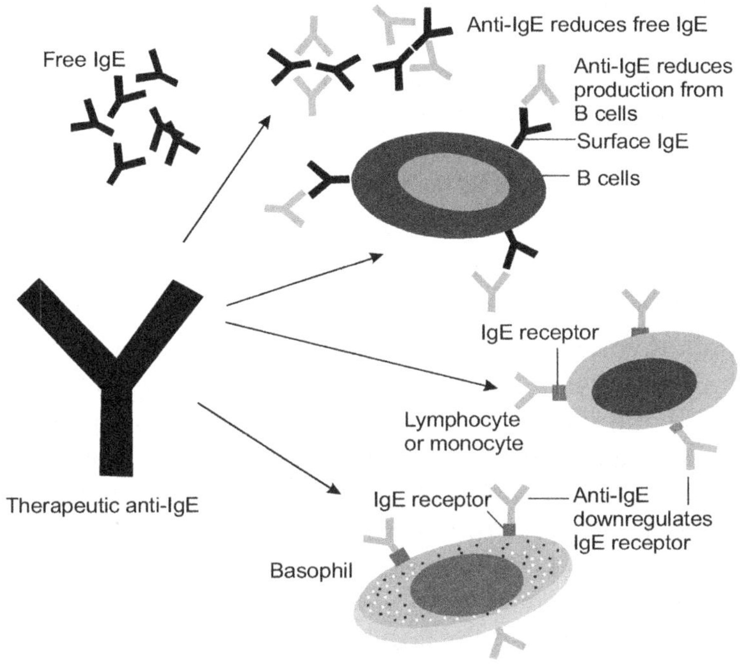

Figure 7.30 Xolair combines with free IgE and IgE bound to receptors present in B cell, monocyte and basophil

Omalizumab is also used to treat atopic dermatitis, allergic rhinitis and food allergies. Anyone suffering from allergies can get a standard anti-IgE shot that will prevent all allergic reactions, no matter what type. At the 2005 AAAAI Annual Meeting in San Antonio, researchers reported the first successful treatment of imported fire ant aphylaxis with Omalizumab.

In a study assessing the immunologic effects of Omalizumab on airway inflammation in allergic asthma patients, the mean percentage of sputum eosinophil count decreased significantly ($p < 0.001$) from 6.6 per cent to 1.7 per cent. This reduction observed in the Omalizumab group is significantly ($p = 0.05$) greater than with placebo (8.5% to 7.0%). This was associated with a significant reduction in the tissue eosinophils, $Fc\varepsilon RI^+$ cells, $CD3^+$, $CD4^+$ and $CD8^+$ T lymphocytes, B lymphocytes, and cells staining for $IL4^+$, but not with improvement in airway hyperresponsiveness to methacholine. Treatment with Omalizumab normalized the number of myeloid dendritic cells during the grass-pollen season.

Omalizumab induces complex changes in the interleukin levels and does not decrease all Th-2-related interleukins. In a study of asthmatics, circulating levels of IL-5, IL-6, IL-8, IL-10, IL-13, and s-ICAM were measured before and after 16 weeks of treatment. The results of the aforementioned study were interesting: IL-13 decreased significantly ($p < 0.01$)); IL-5 and IL-8 decreased in the Omalizumab group compared to baseline; and other circulating mediators remained unchanged.

In approving Omalizumab, FDA considered two 52-week phase-III trials involving 1,071 asthma patients, along with the data from several supportive, safety and efficacy studies. When used as an add-on therapy to inhaled corticosteroids, Omalizumab reduced the mean asthma exacerbations per patient by 33–75 per cent during the stable-steroid phase and by 33–50 per cent during the steroid-reduction phase. In the clinical studies of asthma patients, it improved the peak response, reduced the exacerbations, and improved the quality of life scores. Treatment with intravenous and subcutaneous Omalizumab resulted in a 98–99 per cent reduction in free IgE whereas no reductions were observed following placebo treatment.

This product has been shown to be safe and effective to treat people (12 years of age and older) with moderate-to-severe allergy-related asthma inadequately controlled with inhaled steroid treatments. In these patients, Omalizumab treatment has been shown to decrease the number of asthma exacerbations or the episodes of airway narrowing that result in wheezing, breathlessness, and cough. The product is given as an injection under the skin. It is estimated that a relatively small percentage of these people would be appropriate candidates for this new drug. It is a second-line treatment, recommended only if the first-line treatments have failed.

About 80–85% of Xolair-treated patients had no exacerbations compared to about 70–75% of placebo-treated patients. During the clinical trials, more patients treated with Xolair developed a new or recurrent cancer (0.5%) compared to control patients (0.2%). The sponsor is planning for long-term studies in an attempt to determine whether there is a relationship between Xolair treatment and cancer. A drawback to this new medication is that it is a shot that is given once or twice a month. In addition, it is very expensive, costing the patients about $10,000 per year. If it reduces the hospital visits and hospitalizations, it will likely be a cost-effective treatment though, so hopefully insurance companies will pay for it.

MONOCLONALS IN OSTEOPOROSIS

Osteoporosis is a disease in which bones become fragile and more likely to break. If not prevented or if left untreated, osteoporosis can progress painlessly until a bone breaks. These broken bones, also known as fractures, occur typically in the hip, spine, and wrist. This is more common in women than in men.

Monoclonal AMG-162

AMG-162 (Amgen, Thousand Oaks, CA, USA) is being tested (phase III) in patients with osteoporosis and treatment-induced bone loss (TIBL). This monoclonal is directed against RANKL, a molecule involved in the regulation of bone remodelling. In a single-dose, placebo-controlled study in postmenopausal women, AMG-162 was found to have a dose-dependent and sustained activity in blocking the bone resorption, with no reported serious drug-related adverse events.

EVOLUTION OF MONOCLONALS

It was 11 years before the first monoclonal therapy reached the market (Ortho Biotech's Orthoclone OKT3, Muronomab-CD3) and another 8 years until the FDA approved the second antibody-based therapy (Centocor's ReoPro, abciximab). Thus, 20 years after the discovery of the process to make monoclonals, only two monoclonal-based therapies had reached the market in the United States. In general, FDA-approved monoclonals based on new technologies have emerged between 10 and 12 years after the date they were reported in the scientific literature (Figure 7.31).

Most of the monoclonals that entered clinical studies between 1980 and 1987 were **murine monoclonals**. Mouse-generated monoclonal antibodies, however, were often rejected by patients due to the production of human anti-mouse antibody or HAMA response. This response reduces the effectiveness of the antibody by neutralizing the binding activity and by rapidly clearing the antibody from circulation in the body. The HAMA response can also cause significant toxicities with subsequent administrations of mouse antibodies, and so their use declined drastically between 1987 and 1990s, then gradually dropped to zero in 2003.

From 1985 **chimeric monoclonals** like Rituximab, Basiliximab, Infliximab and Cetuximab were developed and marketed. Although chimeric antibodies are "more human" and theoretically, less likely to trigger an immune reaction, they nonetheless can trigger a human anti-chimera antibody (HACA) response by the human immune system. Scientists then engineered antibodies as **humanized or CDR-grafted** antibodies which contain approximately 5% to 10% mouse protein sequences in the CDR regions. Trastuzumab (herceptin), Palivizumab (Synagis), Daclizumab (Zenopax), Alemtuzumab (Campath), Gemtuzumab (Mylotarg), Vitaxin, Omalizumab (Xolair), Efalizumab (Raptiva) and Bevacizumab (Avastin) are examples of humanized antibody drugs. They revitalized the magic bullet concept of antibody-based immunotherapy.

It is possible to make fully **human antibodies** (100% human protein sequences) by using transgenic mice in which mouse antibody gene is knocked out and effectively replaced with human antibody gene repertoire. A wave of fully human antibodies have been launched from 2006. Future applications of this technology include the potential for creating large transgenic farm animals that can be directly used for the production of therapeutic human-sequence-specific monoclonal or polyclonal antibodies.

BioInvent as well as Morphosys and Cambridge Antibody Technology (CAT) (now Astra Zeneca) patented library technology for antibody fragments. They developed more than 20 billion differentiated monoclonal antibody fragments from their library.

Advances in immunoconjugate technology, complicated immuno-liposomes, bispecific antibody, diabodies, pre-targeted radio-immunotherapy and finally improved antibody cytotoxicity have been gaining more and more clinical experiences. Several monoclonals in development are also much smaller than the full molecule. Thus, smaller molecules, including specific domains, domain combinations, fragments modified with polyethylene glycol (PEG), glycosylated monoclonals, isotype-altered monoclonals, mutation-induced improved monoclonals, domain-deleted monoclonals, alternative binding domains/scaffolds, scFv, only Fab fragments, etc. are currently under clinical investigation as drug candidates. Because of their specificity, functional affinity, or avidity, monoclonals can be achieved by increasing the number of binding sites of the antibody. This is done by means of further protein engineering by increasing the number of variable domains per molecule. This kind of second-generation antibody therapies represent newer immunotherapy strategy.

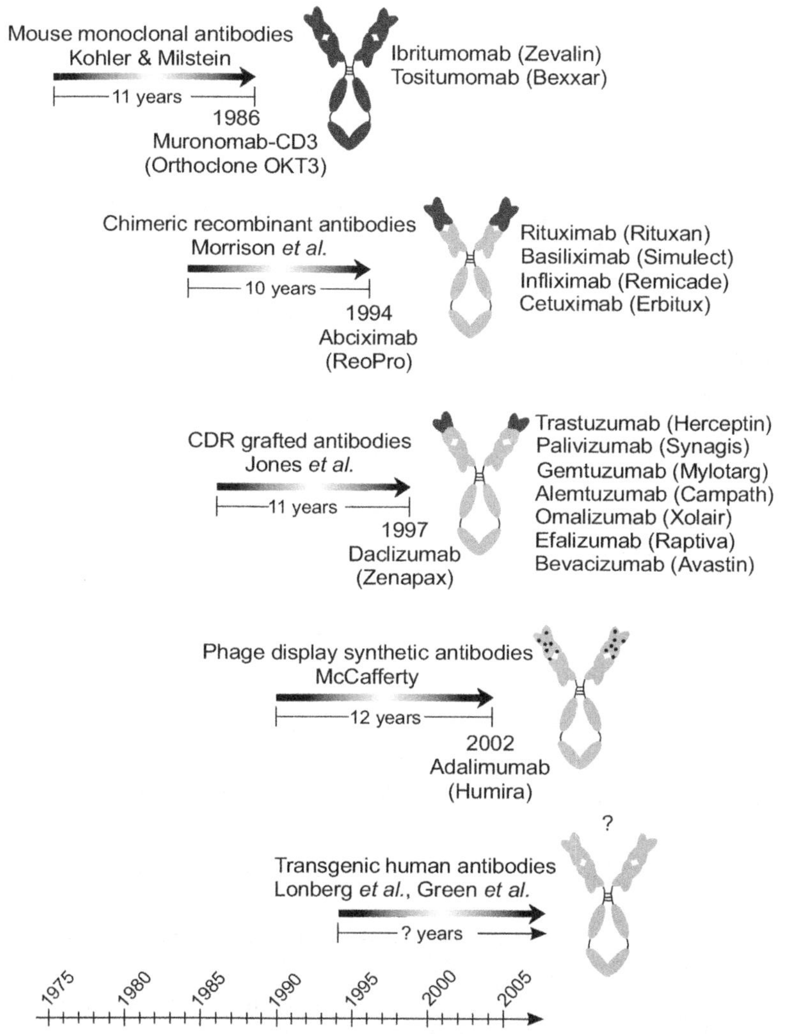

Figure 7.31 Evolution of therapeutic monoclonals (Nils Lonberg, *Nat. Biotech.* 23,2005)

The industries need to continue to evolve towards technology integration and market expansion. Success will depend on strategies targeting shorter development times, higher efficiency rates, innovative molecular engineering, robust intellectual property protection and the development of cost-effective manufacturing. Researchers have to concentrate on improving the characteristics by altering antibodies to boost their efficacy, enable them to more readily penetrate tumours, enhance their ability to stimulate beneficial immune responses.

THE PRESENT AND FUTURE STATUS OF MONOCLONALS

Several thousands of patents and applications are filed in the field of antibody. A general analysis of patents by BioSeeker Group, (February 2004), is shown for second generation antibodies (Table 7.2).

Table 7.2 Patents filed for second-generation antibodies

Nature of monoclonals	Patent
Modification of 'V' region by introducing mutation in the glycosylation sites	PDL
Altered antibody isotype by oligonucleotide synthesis and recombinant DNA technology	Celltech
Mutation in antibody gene to improve function	Patents of Scotgen
ScFv	Enzon and creative biomolecules
Only Fab fragments	Xomas
Improving antibody affinity, avidity and/or half-life	Several patents by PDL and Celltech. Humanized or fully human antibodies–patents filed by Celltech, Genentech, Madarex and Abgenix.
Antibody-effector fusion molecules like CD22-Fv-PE98	Filed by Salvatore Giuliana and immunoconjugate by IDEC.
Altered glycosylation and domain deleted antibody	No patents are found
Non-antigenic toxin-conjugate and fusion protein—a bispecific antibody	By immunomedics for treating leukaemias and lymphomas.
Antibody to treat cancer, viral infection and autoimmune diseases, having TCR-like specificity, yet higher affinity. This is an antibody against MHC-peptide.	Filed by Technion R and D Foundation from Ireland
"Prodrugs of CC-1065 analogs"	Immunogen filed a patent
Ig/B7.1 and Ig/RANTES fusion protein	Filed by Dr. Joseph and D. Roseblaff
PFB4 (Fv)-PE38	By researchers at NCI

Epicyte, a pharmaceutical company has been teamed up with Dow to produce plantibodies from corn that will be formulated either as creams or ointments for mucosal surfaces such as the lips and genitalia, or as orally administered drugs for gastrointestinal or respiratory infections. It has also been applied FDA clearance to begin clinical trials of corn-produced plantibodies to prevent the transmission of herpes simplex and human papilloma virus, which causes genital warts and cervical cancer. The company has also been developing monoclonals that bind to the sperm as possible contraceptives.

Table 7.3 R & D activites of various companies after 2003

Company	Research and developmental activities
Abgenixs	Clinical tests of a fully human antibody against interleukin-8 (IL-8) for inflammatory autoimmune diseases. Clinical trials of Panitumumab targeting EGFR.
Medarex	Designed and engineered bispecific immunotoxin for a diseased cell. Clinical trials of MDX-010 for the melanoma patients. MDX-060 Phase II human monoclonal antibody that targets CD30 for the treatment of CD30-positive lymphomas including CD30-positive Hodgkin's disease, anaplastic large cell lymphoma and other CD30-positive cancers. MDX-070 Phase II human monoclonal antibody that targets prostate-specific membrane antigen (PSMA) for treatment of prostate cancer.
Monsanto Corporation	Phase II trials of transgenic corn plantibodies.
Agracetus	Strain of corn that yields 1.5 g/acre of corn to meet the demands of entire US market for 10,000 cancer patients on just 30 acres. Cultivation of soybeans with humanized monoclonal against gpB of HSV-2. Focus on plantibodies that do not need the carbohydrates.
The Planet Biotechnology	Testing an anti-tooth-decay mouthwash made from transgenic tobacco plants.
The ImClone	Bispecific diabody that would bind simultaneously to both insulin-like growth factor receptor and epidermal growth factor receptor. CDP-791 phase II trials of a PEGylated diFab antibody targets VEGFR-2.
Jon Weidanz	An anti-cancer therapeutic antibody generated against a peptide-MHC receptor logic complex that mimics TCR used for therapy and imaging.

(Contd.)

Table 7.3 (Continued)

Company	Research and developmental activities
Antisoma	Phase III trails of R1549 for ovarian cancer.
Immunomedics	Lymphocide (Epratuzumab)–phase II & III clinical trials for NHL. Cea-Cide (Labetuzumab)–Phase I/II trials for solid tumours. Bispecific antibodies for imaging and treatment of pancreatic cancer. Combination therapy of Epratuzumab with CHOP.
Peregrine	Cotara phase III registration.
Active Biotech	Fab conjugated with Staphylococcal enterotoxin-A (TTS-CD8)
ASCO	Tailored Fab C242.
Seattle Genetics	Antibody conjugate (SGN-35) with synthetic dolastin auristatin E and monomethylauristatin E.
Seattle Genetics	Clinical trials of 2F8-monoclonal against EGFR. Phase III clinical trials of HuMax-CD20 (Ofatumumab) for chronic lymphocytic leukaemia (CLL) and non-Hodgkin's lymphoma. Phase III trials of HuMax-EGRf (Zalutumumab) for head and neck cancer.
Genmab Copenhagen	Clinical trials of AMG-162 for osteoporosis. Phase III trials of Panitumumab (rHuMAb-EGFr) a human monoclonal antibody that targets the epidermal growth factor receptor (EGFR) for colorectal cancer.
Amegen	Phase II clinical trials of Volociximab—an anti-alpha5 beta 1 integrin (a cell adhesion molecule) that inhibits angiogenesis.
Biogen Idec and PDL BioPharma	Phase II clinical trials of anti-CD80 Mab (Galiximab) for non-Hodgkin's B-cell lymphoma. Phase II clinical trials of anti-CD23mab (Lumiliximab) for chronic lymphocytic leukaemia.
Biogen Idec	Phase II clinical trials of Omnitarg (Pertuzumab), a monoclonal antibody for metastatic breast cancer. It is the first in a new class of therapies called HER dimerization inhibitors (HDIs). It blocks the ability of the HER2 receptor to collaborate with other HER receptor family members. Phase II clinical trials of Ocrelizumab a fully humanized anti-CD20 antibody. Phase II clinical trials of a second-generation anti-CD20 for rheumatoid arthritis.
Genentech and Roche	Phase III clinical trials of HuMax-CD-4 (zanolimumab), a human antibody targeting the CD4 receptor. Phase III clinical trials for cutaneous T-cell lymphoma (CTCL), and Phase II for non-cutaneous T-cell lymphoma.

(*Contd.*)

Table 7.3 (Continued)

Company	Research and developmental activities
Serono and Genmab	Phase II clinical trials of Adecatumumab (previously called MT201) a human monoclonal antibody targeting the epithelial cell adhesion molecule (Ep-CAM) for breast and prostate cancer.
Serono	Phase III clinical trials of Actemra (Tocilizumab) (R1569), a humanized anti-IL-6 receptor monoclonal antibody for rheumatoid arthritis and systemic juvenile idiopathic arthritis.
Roche	Phase III clinical trials of OvaRex (Oregovomab) monoclonal antibody that targets CA125, an ovarian cancer tumour marker. OvaRex induces an immune response against CA125, and therefore against ovarian cancer.
United Therapeutics	Phase II clinical trials of TNX-355 a human monoclonal antibody that binds to the CD4 receptor, thus preventing entry of HIV into the cell for treatment of HIV/AIDS.
Tanox (acquired by Genentech 11/06)	Phase I/II clinical trials of MYO-029 a recombinant human antibody that binds to and inhibits the activity of myostatin (growth and differentiation factor 8, or GDF-8) for treatment of muscular dystrophy.
Wyeth	Recombinant human antibody that binds to and inhibits the activity of myostatin (growth and differentiation factor 8, or GDF-8) for treatment of muscular dystrophy

Whether they come from cattle, goats, corn or bioreactors, monoclonal antibodies are set to become a major part of the 21st-century medicine. More monoclonal antibody drugs should be expected on pharmacy shelves in the near future. At present about 175 antibody products are in development. More than 40 new monoclonal antibodies are in phase II or later development, setting innovative technology platforms for a continued and perhaps expanding influx of monoclonal antibody therapeutics.

The monoclonal antibody market for diagnostic and therapeutic antibodies is large and well established. It is one of the fastest growing and most lucrative sectors of the pharmaceutical industry, with exceptional 48.1% growth between 2003 and 2004. The Global Market (BIO016G) from BCC research for monoclonal therapeutic and diagnostic antibodies was worth $31 billion in 2007. This is expected to increase to over $56 billion by 2012, with a **compound average annual growth rate (CAGR)** of 13%.

The market is broken down into applications of therapeutic and diagnostic technologies. Of these, therapeutic technologies have the largest share of the market. At an estimated worldwide market value of $24 billion in 2007, monoclonal antibodies for therapeutic use are already a significant part of effective medical treatment and the negotiations of business deals within the biotechnology and pharmaceutical sectors of the economy. They are poised to reach an estimated value of $47 billion by 2012 for a CAGR of 14%. The diagnostic market for mAb has a 7% CAGR. It encompasses both diagnostic radiopharmaceuticals injected into the body to image areas of interest and immunoassay tests that use antibodies to detect substances of interest. Valued at $6.5 billion in 2007, this segment is expected to be worth $9 billion by the end of 2012 (Table 7.4).

Table 7.4. Global sales of monoclonal therapeutic and diagnostic antibodies 2005–2012 ($ billions)

Technology type	**2005**	**2006**	**2007**	**2012**	**CAGR% 2007–2012**
Therapeutic	13	20	24	47	14
Diagnostic (Antibody-based immunoassay tests and diagnostic radiopharmaceutical)	5.7	6	6.5	9	7
Total	19	26	31	56	13

Source: BCC Research

Oncology products will continue to dominate the market. However, sales of therapeutic products for arthritis, immune and inflammatory disorders (AIID) are forecast to grow strongly and account for 40.1% of the market by 2010. Roche and Genentech dominate the monoclonal market, with a combined market share of 44.9% in 2004, although this is forecast to slip to 35.7% in 2010. Datamonitor identified just 17 biotech companies with direct sales of monoclonal antibodies in 2004, but this figure is expected to more than double to 36 in 2010, as new products and companies flood the market.

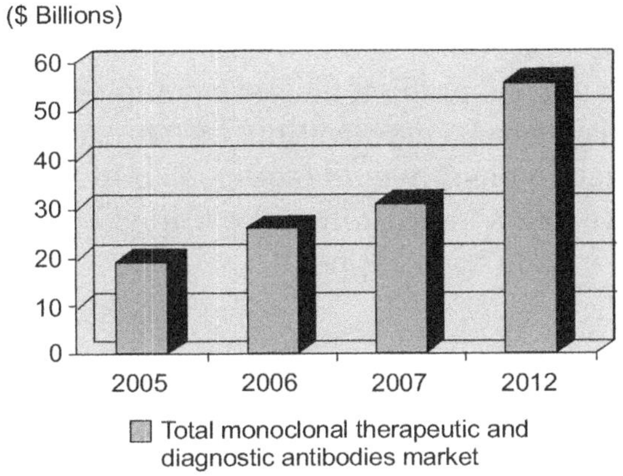

Figure 7.32 Global monoclonal therapeutic and diagnostic antibodies market ($ billions)

The key factor that influences the sales growth in pharmaceutical market over the period 2006–12 is the **generic competition**. Exposure to patent expiries and generic competition will underpin the tepid CAGR in sales of 2.2% forecast out to 2012. Generic competition, known as the **patent cliff**, will act as a notable 'brake' on sales growth and will drive an overall decline in market revenues over the period 2011–12, equal to a year-on-year decline of –7.3%. The 'patent cliff' will be caused by a raft of blockbuster drugs losing patent exclusivity over this short time period.

Part of the problem is financial: banks do not want to lend the hundreds of millions of dollars it takes to build a state-of-the-art monoclonal production facility unless the likelihood that the plant will generate profits is all but guaranteed. Many of them look back the 1980s, when the first generation MAbs of the 1980s attracted large investment and then failed to deliver—and that has meant a lingering bad taste in the mouths of many bankers.

With all these good opportunities and difficulties, biotechnology and pharmaceutical companies might be expected to be ramping up their production lines in anticipation of a big market surge. But worldwide just 10 large-scale antibody plants are now operating.

The gold standard for producing monoclonals from hybridomas relies on enormous tanks called bioreactors. V. Bryan Lawlis, chairman of Diosynth ATP in Cary, N.C., estimates that one giant, 60,000-litre bioreactor plant would be able to (hypothetically) accommodate only four products. Assuming that 100 monoclonals will be on the market by 2010, as analysts predict, Lawlis calculates that the industry will need to build at least 25 new facilities or "we can't satisfy all the needs." Those production plants would require $5 billion or more and between three and five years to be built and certified by the FDA—a prospect no one thinks is going to happen.

REFERENCES

Antibodies and therapy-A hundred years of antibody therapy. (users.path.ox.ac.uk/~scobbold/tig/Swdsp2.gif HMS Beagle (BioMedNet) web pick).

Baert, F., Noman, M., Vermeire, S., Van Assche, G., D' Haens, G., Carbonez, A., Rutgeerts, P. (2003). "Influence of immunogenicity on the long-term efficacy of Infliximab in Crohn's disease." *N. Engl . J. Med.* **348**: 601–608.

BioSeeker Group (2004). Cancer highlights – improved monoclonal on the rise, No.**2**. March. (www.bioseeker.com).

Carter, P. (2001). "Improving the efficacy of antibody-based cancer therapies." *Nat. Rev. Cancer.* **1**(2): 118–29.

Chang, Q., Zhong, Z., Lees A., Pekna, M. and Pirofski, L. (2002). "Structure-function relationships for human antibodies to pneumococcal capsular polysaccharide from transgenic mice with human immunoglobulin loci." *Infect. Immun.* **70**: 4977–4986.

Cone, R.A. and Whaley, K.J. (1994). "Monoclonal antibodies for reproductive health–Part I, Preventing sexual transmission of disease and pregnancy with topically applied antibodies." *Am. J. Reprod. Immunol.* **32**: 114–131.

Co, M.S. and Queen, C. (1991). "Humanized antibodies for therapy." *Nature.* **51**: 501–502.

Corren, J. Ashby, M. and Casale, T.B. (2003). " Omalizumab, a recombinant humanized anti-IgE antibody reduces asthma related emergency room visits and hospitalization in patients with allergic asthma." *J. Allergy. Clin. Immunol.* **111**: 87–90.

Czuczman, M.S. (1999). "Chop plus Rituximab chemo immuno therapy of indolent B-cell lymphoma." *Semin. Oncol.* **26:** 896.

Elias, C. and Heise, L. (1994). "Challenges for the development of female-controlled vaginal microbicides." *AIDS.* **8**: 1–9.

Ezzell Carol. (2001). Magic bullets fly again. *Scientific American*. Oct. 2006. http://www.sciam.com/2001/1001issue/1001ezzell.html.

Feldmann, M., Brennan, F.M. and Maini, R.N. (1996). "The role of cytokine in rheumatoid arthritis." *Ann. Rev. Immunol.* **14**: 397.

Frankel, A.E., Kretman, RJ. and Sausville, E.A. (2000). "Targeted toxins." *Clin. Res.* **6**: 326–334.

Garber, K. (2001). "Biotech industry faces new bottleneck." *Nat. Biotechnol.* Vol. **19**: No. 3, 184–185.

Goldberg, D. (2001). "The role of radiolabelled antibody in the treatment of Non-Hodgkins lymphoma–the coming of ageadioimmunotherapy." *Crit. Rev. Oncol. Hematol.* **39:** 195–201.

Halim, N.S. (2000). "Monoclonal antibodies: A 25-Year Roller Coaster Ride." In: *The Scientist.* Vol. **14,** No. 4, 16.

Halin, C., Zardi, L. and Neri, D. (2001). "Antibody-based targeting of angiogenesis." *News in Physiological Sciences.* Vol. **16,** No. 4, 191–194.

He, Y., Honnen, W.J., Krachmarov, C.P., Burkhart, M. Kayman, S.C., Corvalan, J. and Pinter, A. (2002). "Efficient isolation of novel human monoclonal antibodies with neutralizing activity against HIV-1 from transgenic mice expressing human Ig loci." *J. Immunol.* **169**: 595–605.

Hofmesister, J.K., Cooney, D. and Coggeshall, K.M. (2000). "Clustered CD20 induced apoptosis-src-family kinase, the proximal regulator of tyrosine phosphorylation, calcium influx and caspase-3-dependent apoptosis." *Blood Cells Mol. Dis.* **26**: 133–143. http://www.fda.gov/cder/biologics/biologics table.htm.

Jacques Dantal and Soulillo, J.P. (2005). "Immunosupressive drugs and the risk of cancer after organ transplantation." *N. Engl. J. Med.* **352**: 1371–1373.

Kreitman, R.J., Wilson, W.H., Bergeon, K., Raggio, M., Stevenson, M.A., David, J., Gerald, F. and Pastan, I. (2001). "Efficacy of anti-CD22 recombinant immunotoxin BL22 in chemotherapy resistant hairy cell leukemia." **345**: 241–247.

Kola, I. and Landis, J. (2004). "Can the pharmaceutical industry reduce attrition rates?" *Nat. Rev. Drug Discov.* **3**: 711–715.

Kourbeti, I.S. and Boumpas, DT. (2005). "Biological therapies of autoimmune diseases, Current Drug Targets." *Inflammation and Allergy*. **4**: 41–46.

Mathas, S., Rickers, A., Bommert, K., Dörken, B. and Mapara, M.Y. (2001). "Anti-CD20 and B cell receptor mediated apoptosis–evidence for shared intramolecular signaling pathways." *Cancer Res.* **60**: 7170–7176.

Mukherjee, J., Chios, K., Fishwild, D., Hudson, D., Donnell, S.O., Rich, S.M., Rolfe, A.D. and Tzipori, S. (2002). "Production and characterization of protective human antibodies against Shiga toxin 1." *Infect. Immun.* **70**: 5896–5899.

Mukherjee, J., Chios, K., Fishwild, D., Hudson, D., Donnell, S.O., Rich, S.M., Rolfe, A.D. and Tzipori, S. (2002). "Human Stx2-specific monoclonal antibodies prevent systemic complications of *Escherichia coli* O157:H7 infection." *Infect. Immun.* **70**: 612–619.

Nils Lonberg. (2005). "Human antibodies from transgenic animals." *Nat. Biotechnol.* **23**: 1117–1125.

Radomsky, M.L., Whaley, K.J., Cone, R.A. and Saltzman, W.M. (1992). "Controlled vaginal delivery of antibodies in the mouse." *Biol. Reprod.* **47**: 133–140.

Ross, J.S. and Fletcher, J.A. (1998). "The HER-2/neu oncogene in breast cancer: prognostic factor, predictive factor, and target for therapy." *Stem Cells.* **16**(6): 413–428.

Shan, D., Ledbetter, J.A. and Press, O.W. (1998). "Apoptosis of malignant human B cells by ligation of CD20 with monoclonal antibodies." *Blood.* **91**: 1644–1652.

Sherwood, J.K., Zeitlin, L., Whaley, K.J., Cone, R.A. and Saltzman, W.M. (1996). "Controlled release of antibodies for long-term topical passive immunoprotection of female mice against genital herpes." *Nat. Biotechnol.* **14**: 468–471.

Skov, L., Kragballe, K., Zachariae, C., Obitz, E.R., Holm, E.A., Gregor, B., Jemec, E., Sølvsten, H., Ibsen, H.H., Knudsen, L., Jensen, P., Petersen, J.H., Menné, T. and Baadsgaard, O. (2003). "HuMax-CD4: a fully human monoclonal anti-CD4 antibody for the treatment of psoriasis vulgaris." *Arch. Dermatol.* **139**:1433–1439.

Slamon, D.J., Leylan-Jones, B., Shak, S., Fuchs, H., Paton, V., Bajamonde, A., Fleming, T., Eiermann, W., Wolter, J., Pegram, M., Baselga, J. and Norton, L. (2001). "Use of chemotherapy plus a monoclonal antibody against HER2 for metastatic breast cancer that over expresses HER2." *N. Engl. J. Med.* **344**: 783–792.

Stern, M. and Herrmann, R. (2005). "Overview of monoclonal antibodies in cancer therapy: present and promise." *Critical Reviews in Oncology/ Hematology.* **54**: 11–29.

Subramanian, K.N., Weisman, L.E., Rodes *et al.* (1998). "Safety, tolerance and pharmacokinetics of humanized monoclonal antibody to respiratory Syncytial virus in premature infants and infants with broncho pulmonary dysplasia." *Pediatri. Infec. Dis. J.* **17**: 110–115.

Whaley, K.J., Zeitlin, L., Barratt, R.A., Hoen, T.E. and Cone, R.A. (1994). "Passive immunization of the vagina protects mice against vaginal transmission of genital herpes infections." *J. Infect. Dis.* **169**: 647–649.

Williamson, R., Burioni, R., Sanna, P., Patridge, L., Barbas, C. and Burton, D. (1993). "Human monoclonal antibodies against a plethora from single combinatorial library." *Proc. Natl. Acad. Sci.* USA. **90**: 4141–4145.

Wong, J. (1998). "Monoclonals to hit stride in '98." *Genetic Engineering News.* January, 21.

Zeitlin, L., Whaley, K.J., Sanna, P.P., Moench, T.R., Bastidas, R. and De Logu, A. (1996). "Topically applied human recombinant monoclonal IgG1 antibody and its Fab and F(ab′)2 fragments protect mice from vaginal transmission of HSV-2." *Virology.* **225**: 213–215.

Zeitlin, L. (1996). Doctoral dissertation, Topical methods for preventing genital herpes infection in the mouse. The Johns Hopkins University, Baltimore, MD,

Zeitlin, L., Olmsted, S.S., Moench, T.R., Martinell, B.J., Paradkar, V.M., Queen, C., Cone, R.A. and Whaley (1998). "A humanized antibody produced in transgenic plants for immunoprotection of the vagina against genital herpes." *Nat. Biotechnol.* **16**: 1361–1364.

Zeitlin Larry, Richard, A., Cone and Kelvin, A. Whaley. (1999). "Using monoclonal antibodies to prevent mucosal transmission of epidemic infectious diseases." *Emerging Infectious Diseases.* Vol. **5**. No.1.

Question Bank

I. Choose the correct answer:

1. In the hybridoma technology the marker enzyme responsible for selection of hybridoma cells is
   a. Thymidine kinase
   b. Hypoxanthine guanine phosphoribosyl transferase
   c. Phosphoribosyl 1-pyrophosphatase
   d. Dihydrofolate reductase
   e. Nucleoside phosphorylase

2. Which one of the following statements about monoclonal antibodies is FALSE?
   a. Infliximab is a monoclonal antibody used in the treatment of Crohn's disease.
   b. Herceptin used for breast cancer downregulates HER-2 proteins and their receptors.
   c. Orthoclone OKT3 is the first therapeutic monoclonal used for non-Hodgkin's lymphoma.
   d. sIgA monoclonal is used for the treatment of dental, diarrhoeal, respiratory or vaginal infections.
   e. Omalizumab stops free-serum IgE from attaching to mast cells and prevents inflammation.

3. Select the monoclonal that is less immunogenic to human when it is used as drug.
   a. Mouse monoclonal
   b. Chimeric monoclonal
   c. Humanized monoclonal
   d. Human monoclonal
   e. Catmab

4. A tetroma is the fusion product of
    a. Normal B cell with another normal B cell
    b. Normal cell with another hybridoma cell
    c. Hybridoma cell with another hybridoma cell
    d. Hybridoma cell with T helper or killer cell
    e. Myeloma cell with another hybridoma cell
5. Kohler and Milstein used which of the following medium to select hybridoma cells?
    a. 2 X Pen-Strep medium
    b. DEM medium
    c. RPMI-1640 medium
    d. HAT medium
    e. 8-Azaguanine medium
6. Which of the following is correct about monoclonal antibodies?
    a. Monovalent monospecific
    b. Monovalent bispecific
    c. Monovalent multispecific
    d. Bivalent monospecific
    e. Bivalent bispecific
7. Which one of the following monoclonal is cleared rapidly from the blood during imaging?
    a. IgG
    b. scFv
    c. sIgA
    d. IgE
    e. diabodies
8. Tick the most appropriate statement with regards to hybridoma selection in HAT medium
    a. B cells cannot use exogenous hypoxanthine and hence die.

b. Myeloma cells due to aminopterin cannot synthesize purines by de novo pathway.

c. Myeloma cells due to HGPRT deficiency cannot synthesize purines by salvage pathway.

d. Hybridoma cells utilize HGPRT from B cell after fusion hence survive in the medium.

e. All the above.

9. The mechanism of action of Infliximab in Crohn's disease involves

a. Binding to TNF alpha

b. Blocking TNF alpha receptor

c. Reducing TNF alpha production

d. Reducing TNF alpha secretion

e. Increasing hepatic metabolism of TNF alpha

10. The major facility required for hybridoma technology includes

a. Animal room, cell culture room, well plates, RIA facility

b. Animal room, cell culture room, centrifuge, inverted microscope

c. Animal room, cell culture room, centrifuge, CO_2 incubator, ELISA reader

d. Cell culture room, centrifuge, inverted microscope, ELISA reader

e. Animal room, cell culture room, facility for cryopreservation

11. You wish to produce a monoclonal drug for an epidemic disease. Among the following, which method would you use for production of the drug?

a. Ascite method

b. T-flasks and roller bottles

c. Membrane-based and matrix-based culture systems

d. High cell density bioreactors

e. Unit-process fermenter system

12. Which one of the following statements is **TRUE** about monoclonal antibodies?

    a. To make fully humanized or human antibodies, mouse antibodies are cleaved enzymatically.

    b. Minibody can be produced by linking two scFv by CH_3 domains and a hinge region peptide.

    c. Pegylated antibodies show decreased resistance to proteolysis and circulating half-life.

    d. Pepbodies are the antibody fragments with fused peptide that may stay longer in tissue.

    e. The first monoclonal successfully developed to combat an infectious disease is Muronomab.

13. 'Primatized antibodies' are antibodies produced

    a. during early infection

    b. prior to humanization

    c. prior to antibody engineering

    d. from monkey

    e. from primary cultures

14. In hybridoma technology, HAT media is a

    a. primary medium

    b. minimal medium

    c. supplementary medium

    d. specific medium

    e. selection medium

15. Transfectomas are obtained by

    a. Transfecting hybridoma cells with another spleen cell

    b. Transfecting Ig gene into prokaryotic hosts

    c. Using M13 phage library and obtaining the desired antibody by screening the library

    d. Transfecting lymphoid cells or injecting with desired Ig genes and obviating the cell fusion

    e. Supplementing antibody gene through gene therapy

16. A radiologist has to image a patient with recurrent and metastatic colorectal cancer. He has to select a monoclonal for imaging. Tick the one he will select from the following.

    a. Prostoscint PSMA (Capromab pendetide)
    b. CEA-Scan (Arcitumomab)
    c. Verluma (Nofetumomab)
    d. Oncoscint CR/OV (Satumomab pendetide)
    e. Myoscint (Imciromab)

17. Monoclonal antibodies are used extensively for the following **except**

    a. Monoclonals are essential reagents for isolation, identification and cellular localization of specific gene products, the transcription factors and macromolecular structures.
    b. It is used as emergency contraceptive to prevent the onset of pregnancy.
    c. The maximum drug residue limit in food products is detected using monoclonals against antibiotics or toxic products.
    d. Monoclonal antibodies can be used to prevent melanoma, prostate, breast, pancreatic and lung cancers.
    e. They can be used to increase the memory power and reduce the ageing process.

18. A pharmacist is visiting an asthma specialist. Among various monoclonal drugs which drug catalogue will he hand over to that medical practitioner?

    a. Remicade
    b. Soliris
    c. Xolair
    d. Reopro
    e. Zenapax

19. The monoclonal that is more often used by a cardiologist is

    a. Reopro
    b. Rituxan

c. Herceptin

d. Mylotarg

e. Remicade

20. Four patients suffering from 1) non-Hodgkin's lymphoma, 2) colorectal cancer 3) rheumatoid arthritis and 4) psoriasis were approaching a general medical practitioner. From the list of following FDA-approved monoclonal drugs which drug will he prescribe for each patient?

a. Muronomab-CD3 (Orthoclone-OKT3)

b. Bevacizumab (Avastin)

c. Trastuzumab (Herceptin)

d. Infliximab (Remicade)

e. Omalizumab (Xolair)

f. Palivizumab (Synagis)

g. Rituximab (Rituxan)

h. Efalizumab (Raptiva)

Patient 1 -

Patient 2 -

Patient 3 -

Patient 4 -

II. Match and choose the correct answer:

1.

Monoclonal drug	**Target antigen**
1. Avastin	a. ErbB2
2. Rituxan	b. CD52
3. Herceptin	c. VEGF
4. Campath	d. CD20

A. 1d, 2c, 3b, 4a B. 1c, 2d, 3a, 4b

C. 1b, 2a, 3d, 4c D. 1c, 2a, 3d, 4b

2.

Monoclonal drug	**Disease treated**
1. Xolair	a. Cardiovascular disease
2. Zevalin	b. Asthma

3. Synagis	c. Non-Hodgkin lymphoma
4. Reopro	d. RSV infection

A. 1b, 2c, 3d, 4a B. 1b, 2a, 3d, 4c
C. 1c, 2d, 3a, 4b D. 1d, 2c, 3b, 4a

3. | **Monoclonal drug** | **Type of monoclonal** |
|---|---|
| 1. Orthoclone-OKT_3 | a. Radiolabelled monoclonal |
| 2. Herceptin | b. Conjugated immunotoxin |
| 3. Mylotarg | c. Murine monoclonal |
| 4. Bexxar | d. Humanized monoclonal |

A. 1b, 2c, 3d, 4a B. 1c, 2d, 3b, 4a
C. 1d, 2c, 3b, 4a D. 1b, 2a, 3d, 4c

4. | **Monoclonal drug** | **Trade name** |
|---|---|
| 1. Abciximab | a. Remicade |
| 2. Palivizumab | b. Reopro |
| 3. Infliximab | c. Xolair |
| 4. Omalizumab | d. Synagis |

A. 1c, 2d, 3b, 4a B. 1d, 2c, 3b, 4a
C. 1b, 2d, 3a, 4c D. 1b, 2c, 3d, 4a

5. | **Monoclonal drug** | **Application** |
|---|---|
| 1. $38C_2$ | a. to test animal feed |
| 2. Prostoscint PSMA | b. to detect heart diseaes |
| 3. Toxiklon | c. for industrial catalysis |
| 4. Myoscint | d. to detect prostate cancer |

A. 1b, 2c, 3d, 4a B. 1c, 2d, 3b, 4a
C. 1b, 2d, 3a, 4c D. 1c, 2d, 3a, 4b

6. | **Derivatives of monoclonal** | **Method of production** |
|---|---|
| 1. Diabodies | a. papain digestion |
| 2. scFv | b. pepsin digestion |
| 3. Fab | c. C-terminus of V_L is linked to N-terminus of the V_H by short peptide |

4. Fc

d. C-terminus of V_L is linked to N-terminus of the V_H by very short peptide

A. 1d, 2c, 3a, 4b   B. 1c, 2d, 3b, 4a

C. 1b, 2d, 3a, 4c   D. 1b, 2c, 3d, 4a

III. Read the following assertion and reason statements and write the answer as **A** or **B** or **C** or **D**

A. Assertion and reason are true statements, and the reason is the appropriate explanation of assertion.

B. Assertion and reason are true statements but reason is not the appropriate explanation for assertion.

C. Assertion is the correct statement and reason is wrong statement.

D. Both assertion and reason are wrong statements.

1. **Assertion** Use of mouse-generated monoclonal antibodies declined drastically after 1987.

   Reason Mouse monoclonals cause HAMA response that is harmful to the patient.

   Answer: (   )

2. **Assertion** Antibodies that act as angiogenesis inhibitors can be used either in diagnosis or treatment of cancer.

   Reason Angiogenesis is a rare phenomenon in healthy adults. It is a characteristic feature of aggressive solid tumours.

   Answer: (   )

3. **Assertion** Topically applied monoclonals show less immunogenecity with no major adverse side effect.

   Reason Antibodies delivered to the lumen of a mucosal surface have minimal interaction with circulating immune cells.

   Answer: (   )

4. **Assertion** It is difficult to propagate and harvest transgenic plants.

Reason Transgenic seeds cannot be used as a long term storage vehicles.

Answer: ( )

5. **Assertion** Custom-made enzymes are named as "abzymes".

Reason They are usually artificial constructs, not found in normal humans.

Answer: ( )

IV. Write in one or two sentences:

1. Define monoclonal antibody.
2. Mention the advantages of hybridoma technology.
3. Write down the disadvantages of hybridoma technology.
4. Mention the two requirements of a myeloma cell that is used for hybridoma technology.
5. What chemical is used for fusion in hybridoma technology?
6. Expand HAT medium.
7. What is HGPRT?
8. What is ascite?
9. When will the immune system show HAMA response?
10. How is scFv produced?
11. Define diabodies.
12. What is catmab?
13. Mention the significance of catmab.
14. Expand ADEPT.
15. What is prodrug therapy?
16. Name the vector used in phage display.
17. How are minibodies produced?
18. What are bispecific diabodies?
19. Write the significance of pegylated antibodies.
20. Name an immunotoxin.
21. How are aglycosylated antibodies obtained?
22. Give an example of abzyme.
23. How does anti-cocaine abzyme work?

24. Define chimeric monoclonal.
25. How can a humanized monoclonal be produced?
26. Write down the significance of immunoglobulin transgene in SCID mouse.
27. Define plantibodies.
28. What are intrabodies?
29. Name the first monoclonal drug tried in human beings.
30. Differentiate between scFv and diabodies.
31. Name the monoclonal that inhibits angiogenesis of tumour.
32. What is phytopharming?
33. What are immunosensors?
34. Distinguish between a monoclonal and a polyclonal antibody.
35. Differentiate between HAMA and HACA response.
36. What is immunoliposome?
37. To which antigen is Rituximab targeted?
38. What is the target antigen of Raptiva?
39. Name conjugated radioactive label in Bexxar.
40. What type of monoclonal is Zevalin?
41. What is the speciality of Mylotarg?
42. Name the recombinant tailored antibody targeting EGFR found on the surface of malignant cells of advanced colorectal cancer.
43. Mention the target specificity of Basiliximab or Simulect.
44. Name the blockbuster monoclonal antibody drugs.
45. What is the significance of anti-fibrin monoclonal coupled with tissue-specific plasminogen activator?
46. What is CAT-3888?
47. Name the world's first commercially available catalytic antibody.
48. Name the monoclonal drug that reduces the lesions in multiple sclerosis.
49. What monoclonal drug is used for psoriasis?
50. How is cerebral thrombus treated using monoclonal?

V. Write short notes on:

1. Why are monoclonal antibodies advantageous over polyclonal antibodies? List the disadvantages of polyclonals.
2. List the problems associated with obtaining human monoclonals by conventional hybridoma technique.
3. Write down the principle behind HAT selection.
4. In spite of the prohibition of ascite method, it is still under use in many laboratories. Why?
5. With an illustration, explain the production of chimeric human-mouse monoclonal by recombinant DNA technology.
6. How are antibodies produced from combinatorial library or phage library?
7. "Transgenic plants represent an economical alternative to fermentation-based production systems." Discuss.
8. Discuss the problems of producing of monoclonals through transgenic plants and animals.
9. With an illustration explain the strategy for production of secretory antibodies in plants.
10. Explain the method and advantages of transgenic seeds as bioreactor.
11. "Transgenic animal technology could be useful for generating new human sequence monoclonals starting from unrearranged, germ line-configuration transgenes." Discuss.
12. "Transgenic chickens should be a near perfect pharmaceutical bioreactor for making large amounts of pure recombinant monoclonals." Discuss.
13. How can therapeutic monoclonals be made in the milk of transgenic animals? Discuss the safety concerns.
14. Write briefly about monoclonal based gene therapy.
15. How does FACS utilize monoclonal antibodies?
16. Explain monoclonal based ELISA.
17. Describe the principle behind the pregnancy test.
18. Enumerate the types of immunodiagnostic assays that use monoclonals.
19. Explain how monoclonals are used in western blotting.

20. Discuss the industrial applications of monoclonals.
21. Write an account on prodrug therapy.
22. Why are monoclonal antibodies hailed as "magic bullets"?
23. With specific examples write how immunotoxin helps in killing target cells.
24. Write notes on different types of immunosensors.
25. How is anti-idiotypic vaccine produced?
26. Explain how monoclonals help in purification of proteins.
27. "Monoclonals are indispensable tools in research." Discuss.
28. Explain briefly the applications of monoclonals in environmental protection.
29. Give an account on conjugated monoclonals.
30. Enumerate the obstacles to successful monoclonal cancer therapy.
31. Write an account on monoclonals available for infectious diseases.
32. Briefly explain the mechanism behind Rituximab treatment.
33. How do Herceptin and Avastin cure breast and colorectal cancers respectively?
34. Discuss the mode of action of any two monoclonal drugs that are used for treatment of autoimmune diseases.
35. Discuss briefly the need and advantages of development of mucosal antibodies for most of the infectious diseases.
36. Describe the mechanism of target cell killing of Cetuximab or Erbitux.
37. Explain the mechanism of monoclonal drugs used as immune suppressive drugs.
38. Explain how there is a permanent cure to allergy through monoclonal drugs.
39. How can a blood clot be identified and cured by monoclonals?
40. Discuss the use of Orthoclone OKT3 as a immunosuppressive agent.
41. Comment on the evolution of therapeutic monoclonal antibodies.

VI. Give detailed answers:

1. Describe the steps involved in the production of monoclonal antibodies.
2. Critically comment on the *in vitro* production of antibodies.
3. Write any two techniques by which a library of different antibodies is produced.
4. Explain the various methods of antibody engineering.
5. Explain briefly monoclonal based cancer therapy.
6. Explain how monoclonals are used to detect cancer and other diseases.
7. How are monoclonals used as "target-seeking missiles"? Explain this with any three examples.
8. Enumerate the methods used in genetic engineering of monoclonals.
9. Critically discuss the production of human monoclonals in plants.
10. With diagrams, enumerate the types of monoclonals obtained by protein engineering.
11. Explain how monoclonals are used for diagnosis and treatment of cancer.
12. How are monoclonals used in immunoassay and immunohistochemistry?
13. List any five FDA-approved monoclonal drugs and discuss their mode of action.
14. Write down the mechanism of monoclonals that are used to cure autoimmune disease.
15. Explain the methods by which monoclonals are produced from transgenic animals.
16. Briefly discuss the applications of monoclonal antibodies.
17. Write any five medical applications of monoclonal antibodies with examples.
18. Write an account on abzymes.
19. Critically discuss the methods, advantages, problems and cost of production of plantibodies.
20. Explain how transgenic animals are useful in the production of human therapeutic monoclonals.

ANSWERS

I. Choose the correct answer:

1. b	2. c	3. d	4. c	5. d	6. d	7. b
8. e	9. a	10. c	11. e	12. b	13. d	14. e
15. d	16. b	17. e	18. c	19. a		

20. Patient 1 - g

    Patient 2 - b

    Patient 3 - d

    Patient 4 - h

II. Match and choose:

1. B 2. A 3. B 4. C 5. D 6. A

III. Assertion and reason:

1. A 2. A 3. B 4. D 5. C

Glossary

Abciximab A chimeric monoclonal directed against platelet glycoprotein IIb/IIIa receptor. It inhibits the clumping of platelets by binding the receptors on their surface, normally linked by fibrinogen, preventing fibrinogen and VW factor from initiating platelet aggregation.

Adalimumab A fully human monoclonal against TNF-α that binds with both free TNF-α trimers and membrane-bound cytokines and neutralizes the pathophysiological action of TNF-α.

ADCC (Antibody-dependent cell-mediated cytotoxicity) A phenomenon in which target cells, coated with antibody, are destroyed by specialized non-specific killer cells (NK cells, neutrophils and macrophages) which bear Fc receptors of the coating antibody.

Adjuvant The chemicals that increase non-specifically the immune response to an antigen and hence given along with vaccines or immunogens (e.g. Freund's adjuvant, alum, bacterial LPS).

Affinity A measure of binding strength of a single epitope with a monovalent paratope. It is represented quantitatively by the affinity constant K_a.

Agglutination The clumping or aggregation of particulate antigens by antibodies. Agglutination is shown by red blood cells as well as by bacterial cells or inert particles coated with antigen.

Aglycosylated antibody The antibody with no carbohydrate attachment site. It is obtained by substituting the Asn residue at position 297 by site-directed mutagenesis. Such antibodies failed to exhibit ADCC activity, but a significant level of CDC activity was retained.

Alemtuzumab *See* Campath.

Allergen Non-pathogenic antigens responsible for producing IgE-mediated immediate type I hypersensitivity allergic reactions.

Allergic rhinitis An inflammation of the nasal mucosa often due to an allergic reaction to pollen, dust or other airborne allergens.

Allergy Immune reactions to non-pathogenic antigens, which lead to inflammation and cause deleterious effects in the host. The inflammatory mediators are released by mast cell degranulation.

Angiogenesis Formation of new blood vessels during tumour growth.

Ankylosing spondylitis An autoimmune disease resulting in inflammation in the spinal column.

Antibody A serum glycoprotein formed in response to antigenic stimulation. It is highly specific in binding to the immunizing antigen. It is provided with two heavy chains connected by disulphide bonds and two light chains connected to the heavy chains by disulphide bonds. It has two paratopes of same specificity. It is produced by plasma cells which are activated B cells.

Antibody chips Miniaturized substrates on to which a large number of antibody molecules are attached with high density and in a defined microarray.

Antibody resource page A guide designed by the scientists for the scientists to find companies that sell catalog antibodies and custom monoclonal and polyclonal antibodies.

Antibody titre Level of antibodies in circulating blood.

Antigen A foreign material that evokes a specific immune response; a material used for immunization. Antigens may also be immunogens if they are able to trigger an immune response.

Antigen-binding site A part of an immunoglobulin molecule also called as paratope or complementarity determining region (CDR) that binds to the epitope of a specific antigen.

Antimicrobial agents A general term for the drugs, chemicals, or other substances that either kill or slow down the growth of bacteria (antibacterial), viruses (antiviral), fungi (antifungal) or parasites (antiparisitic).

Antiserum (plural: **antisera**) The fluid component of clotted blood from an immune individual, which contains a heterogeneous collection of antibodies against the immunization molecule. Such antibodies bind with the antigen used for immunization. Each has its own structure, its own epitope on the antigen, and its own set of cross-reactions. This heterogeneity makes each antiserum unique.

Apoplasm The free spaces in plant tissue; includes space within the cell wall and intracellular spaces.

Apoptosis A form of programmed cell death caused by the activation of caspases-3 in mitochondria leading to the fragmentation of DNA and formation of apoptotic bodies.

Ascite fluids Fluid that has accumulated in the abdomen.

Ascites production A method by which monoclonal antibodies are produced in mouse abdominal fluid by intraperitonially injecting the hybridoma cells.

Asthma A disease of the lungs characterized by a reversible airway obstruction, and inflammation of the airway with prominent eosinophil participation. Some cases of asthma are allergic and are mediated, in part, by IgE antibody to environmental allergens. Other cases are provoked by non-allergic factors.

ATCC Expanded as American Type Culture Collection, this is a Global Bioresource Center that stores microbial cultures and distributes biological materials such as cell lines, bacteria, animal and plant viruses and antisera.

Atopic dermatitis A chronic, itchy inflammation of the upper layers of the skin.

Autoimmunity The immune response to "self" antigens or self tissues or components which may have pathological consequences leading to autoimmune diseases.

Avastin *See* Bevacizumab.

Avidin–biotin ELISA An assay using biotinylated antibodies that react with avidin or streptavidin that is labelled with enzymes or fluorescent substances.

Avidity The summation of multiple affinities, for example, when a polyvalent antibody binds to a polyvalent antigen.

B cells Lymphocytes which develop in the bone marrow; these cells carry immunoglobulin and class II MHC antigens on their surfaces and binds to antigen. They are the precursors of antibody-forming plasma cells.

Bacteraemia A condition where the bacteria multiply in blood.

Basiliximab *See* Simulect.

Behcet's syndrome A chronic condition which happens due to disturbances in the body's immune system. It becomes overactive and produces unpredictable outbreaks of unwanted and exaggerated inflammation. This extra inflammation affects the small blood vessels. As a result symptoms occur wherever there is a patch of inflammation, and can be anywhere where there is blood supply.

Bevacizumab The antibody against vascular endothelial growth factor (VEGF) which inhibits angiogenesis and maintenance of existing tumour vessels. It is given in combination with intravenous 5-fluorouracil-based chemotherapy for metastatic carcinoma of the colon or rectum.

Bexxar An anti-neoplastic radioimmunotherapeutic monoclonal, labelled with ^{131}I. It is a murine IgG2a lambda monoclonal directed against CD20 antigen of B lymphocytes and is used for the treatment of non-Hodgkin's lymphoma.

Bioreactors A small or large vessel used to culture cells and tissues.

Biotinylated antibodies Antibodies conjugated to biotin that has high affinity to avidin or streptavidin.

Bivalent antibody An antibody with two similar antigen-binding specificities.

Bone marrow transplantation A procedure used to treat both non-neoplastic and neoplastic conditions not amenable to other forms of therapy.

Booster injections A subsequent antigen (vaccine) dose given to enhance the immune response to the original antigens.

Campath A recombinant DNA-derived IgG-1 kappa humanized monoclonal targeted against CD52, an antigen expressed by eosinophils and monocytes. It works in a way entirely different from chemotherapy for the treatment of patients with B cell lymphoma.

Campath-1H An experimental drug used for treatment of multiple sclerosis, organ transplant rejection and several types of leukaemia. It works by destroying the body's T cells which are believed to be responsible for initiating the destructive process seen in multiple sclerosis.

Carcinogens The chemicals that induce cancer.

CD3 A part of the T cell (antigen) receptor complex. It transduces the antigen binding signal outside the plasma membrane into chemical signals (phosphorylation) in the cytoplasm.

CD4 A co-receptor on helper T cells that binds with class II MHC and that participates in T-cell activation by antigen.

CD antigen Membrane proteins on immune cells that allow for their identification and isolation by monoclonal antibodies. All monoclonal antibodies that react with the same membrane molecules are grouped into a common cluster of differentiation.

CDC Complement-dependent cytotoxicity.

cDNA Complementary DNA sequences obtained from the mRNA by reverse transcription.

CEDIA Cloned enzyme donor immunoassay.

Cell line Cells which can be cloned and propagated indefinitely in tissue culture.

Cell-mediated cytotoxicity (CMC) Killing (lysis) of a target cell by an effector lymphocyte.

Cell-mediated immunity (CMI) Immune reactions mediated by cytotoxic T cells in contrast to humoral immunity, which is antibody-mediated. It is also referred to as delayed-type hypersensitivity.

Cetuximab *See* Erbitux.

Chaperones Diverse family of proteins involved in the assembly and transmembrane translocation of other proteins, and seem to function by stabilizing the folding of proteins.

Chemotherapy A type of treatment for cancer that is given either orally or by infusion into a vein. Chemotherapy kills the cancer cells by interfering with the tumour cell's ability to grow and reproduce. Because chemotherapy drugs travel throughout the whole body, they can also affect the normal cells.

Chimeric antibody The antibody encoded by genes from more than one species, usually with variable region from mouse genes and constant regions from human genes.

Class I and II MHC molecules Proteins encoded by genes in the major histocompatibility complex. Class I MHC molecules are heterodimeric membrane proteins. They are expressed by all nucleated cells and present the antigen to CD8 T cells. Class II MHC molecules are expressed by antigen-presenting cells and present the antigen to CD4 T cells.

Clinical trial A research study conducted with people and designed to evaluate the safety and effectiveness of a new drug, usually in comparison with a standard treatment. If a drug is proven to work well in a clinical trial, it may become a new therapy that can help many people.

Clone Many identical copies of a gene or a cell or an organism, produced from a single precursor.

Cloning A process of making copies of a specific DNA or a gene or genetically identical organism.

Combinatorial joining The joining of segments of DNA to essentially generate a new genetic information, as occurs with Ig genes during the development of B cells. Combinatorial joining allows multiple opportunities for 2 sets of genes to combine in different ways.

Combinatorial library A large collection of antibodies that are formed by a random combination of H and L chains. Since the heavy chain sequences contain only the variable region and the first constant domain, the antibody that forms is called Fab and they are screened by specific antigen.

Complementarity-determining regions (CDRs) Parts of immunoglobulins that determine their specificity and make contact with specific antigen. They are the most variable parts of the molecule and contribute to the diversity of these molecules. There are three such regions (CDR1, CDR2, and CDR3) in each V domains of light and heavy chain.

Complement components An enzymatic system of serum proteins triggered by the classical and alternative pathways, resulting in target cell lysis, phagocytosis, opsonization and chemotaxis.

Complement receptor A receptor found on erythrocytes, lymphocytes, neutrophils, monocytes and macrophages that binds with the three constant domains of antibody.

Constant region (C region) The invariant carboxyl-terminal portion of an antibody molecule, distinct from the variable region that is at the amino-terminal of the chain. The sequence of amino acids in the constant region determines the isotype (α, μ, ε, γ, and α) of heavy chain and the type (κ and λ) of light chain.

Crohn's disease A chronic form of inflammatory bowel disease that causes severe irritation in the gastrointestinal tract.

Cross-reactivity The ability of an antibody, specific for one antigen, to react with a second unrelated antigen; a measure of molecular relatedness between two different antigenic substances, e.g. some antibodies to *Streptococcus pyogenes* bind to human heart tissue, resulting in rheumatic heart disease.

Cryopreservation Rapid freezing in liquid or vapour nitrogen at –196°C to preserve cells for future use.

Cytokines Soluble low-molecular-weight proteins secreted by cells, which have a variety of effects on other cells regulating the intensity and duration of immune response, e.g. interleukin 1 (IL-1).

Cytotoxic (Cytolytic) T cell The cell that kills the target cells bearing appropriate antigen within the groove of an MHC class I molecule that is identical to that of the T cell.

Daclizumab *See* Zenapax.

Database A collection of information organized into interrelated tables of data and specifications of data objects.

Desensitization A procedure in which allergic individuals are exposed to increasing doses of allergen with the goal of inhibiting their allergic reactions.

Dioxins The organic compounds that have two linked benzene rings, e.g. PCDD (polychlorinated dibenzodioxin), PCDF (polychlorinated dibenzofuran) and PCB (polychlorinated biphenols).

E5 monoclonal Both human and mouse monoclonals of IgM type, called E5, bind to endotoxin. It reduces the mortality and improves the outcome of multiorgan failure in patients with refractory shock.

EBV Epstein Barr Virus; a virus that usually infects the B cells.

Edrecolomab Monoclonal antibody targeted to the 17-1A antigen seen in colon and rectal cancer.

Efalizumab A humanized monoclonal for treatment of moderate-to-severe plaque psoriasis. It blocks one of these receptors—leucocyte function antigen-1 (LFA-1) or CD11a on the lymphocyte. By blocking adhesion, it selectively and reversibly blocks the activation, reactivation and trafficking of T cells, that lead to the development of psoriasis.

Effector cells Lymphocytes that can mediate the removal of pathogens or antigens from the body without the need for further differentiation. Effectors are distinct from naive lymphocytes, which must proliferate and differentiate before they can mediate effector functions. They are also distinct from memory cells which must differentiate and sometimes proliferate before they become effector cells.

Electrochemical immunosensors Potentiometric immunosensors based on the change in potential that results when antibody is immobilized on an electrode and its antigen (sample) binds to it.

ELISA Enzyme-linked immunosorbent assay. An assay which detects antigen–antibody binding using antibody complexed with an enzyme that forms a coloured product from a colourless substrate.

ELISPOT An assay used to detect secretions of cells like cytokines; an adaptation of ELISA. The cells are placed over antibodies or antigens attached to a plastic surface. The antigen or antibody traps the secreted products of the cells, which can then be detected by using an enzyme-coupled antibody that cleaves a colourless substrate to make a localized coloured spot.

EMIT Enzyme multiplied immunotechnique, a homogeneous assay in which there is no need to separate bound and free antibody; it is used to detect small molecules like drugs.

Enbrel *See* Etanercept.

Endotoxin A toxin which is part of the bacterial membrane structure, e.g. lipopolysaccharide (LPS) in the gram-negative bacteria.

Enzyme-linked immunosorbent assay *See* ELISA.

Epitope An alternative term for antigenic determinant.

Erbitux A recombinant tailored antibody targeting epidermal growth factor receptor (EGFR) found on the surface of malignant cells of advanced colorectal cancer. It is composed of the Fv region of murine anti-EGFR antibody with human IgG1 heavy and kappa light chain constant regions.

Etanercept A monoclonal against tumour necrosis factor (TNF), used for the treatment of rheumatoid arthritis, polyarticular-course juvenile rheumatoid arthritis, ankylosing spondylitis, psoriatic arthritis and psoriasis.

Exacerbation The sudden worsening of a disease or its symptoms.

F(ab´)2 A fragment of an antibody containing two antigen-binding sites generated by cleavage of the antibody molecule with the enzyme pepsin which cuts at the hinge region C-terminally to the inter-H-chain disulphide bond.

Fab The fragment of an antibody containing one antigen-binding site, generated by cleavage of the antibody with the enzyme papain, which cuts at the hinge region N-terminally to the inter-H-chain disulphide bond and generates two Fab fragments from one antibody molecule.

FACS Fluorescence-activated cell sorter; equipment used to count and separate leucocytes labelled with fluorescent-tagged antibodies to cell-surface molecules.

Fc The fragment (capable of crystallization) of antibody without antigen-binding sites, generated by cleavage with papain; the Fc fragment contains the C-terminal domains of the heavy immunoglobulin chains.

Fc receptor (FcR) A receptor on many types of cell surfaces with specific binding affinity for the Fc portion of an antibody molecule.

Feeder cells Cells used as a substitute for conditioned medium, to improve the plating efficiency of hybridoma cells during single-cell cloning. Feeder cells secrete growth factors and are normally prepared from splenocytes, macrophages, thymocytes, or fibroblasts.

Fluorescent antibody An antibody coupled with a fluorescent dye, used to detect antigen on cells, tissues, or microorganisms with a UV-fluorescent microscope.

Framework region Regions of the antibody molecule which do not bind antigen but whose structure allows for folding of the CDRs so all the contact regions bind antigen.

Freund's complete adjuvant A water-in-oil emulsion that contains an immunogen, an emulsifying agent, and killed mycobacteria, which enhances the immune response to the immunogen after injection: It is termed "incomplete" Freund's adjuvant if mycobacteria are not included.

Fv antibody An antibody molecule composed of variable regions from light and heavy chain connected by disulphide bonds and without Fc.

Gene therapy Treating a disease by replacing, manipulating or supplementing non-functional genes.

Gentuzumab ozogamicin *See* Mylotarg.

Glycosylated antibodies The antibodies with human sugar structures.

Glycosylation The modification of a protein by adding sugar molecules to particular amino acids in the protein.

Graft rejection An immune mechanism that kills transplanted non-self (graft) tissue.

Half-life of antibody The time required for half the amount of antibody to be eliminated from the body.

HAMA (Human Anti-Mouse Antibody) response Antibodies made by humans against mouse monoclonal antibodies that are injected to treat cancer or autoimmunity.

Haplotype All genes inherited from one parent; half of one's genome.

Hapten A small molecule which is immunogenic only when covalently linked to a carrier molecule.

HAT medium A selection medium to select fused hybridoma cells containing hypoxanthine and thymidine, the precursors of nucleic acid, and aminopterin, an antibiotic.

HCG Human chorionic gonadotrophin secreted by chorion, even from the 10th day of pregnancy and it is secreted out along with the urine. Antibodies against this hormone are used to test the pregnancy.

Heavy chain (H chain) The larger of the two types of chains that comprise a normal immunoglobulin.

Herceptin A chimeric antibody; specific binding of this to the outward-facing part of the HER2 receptor appears to prevent the transmission of the permanent growth and division signal to the cell nucleus.

HGPRT Hypoxanthine-guanine phosphoribosyl transferase, an enzyme involved in the synthesis of nucleic acids by salvage pathway.

Hinge region A flexible, open segment of an antibody molecule that allows bending of the molecule. The hinge region is located between Fab and Fc and is susceptible to enzymatic cleavage.

Histocompatibility Literally, it is the ability of tissues to get along; in immunology, it means an identity in all transplantation antigens. These antigens, in turn, are collectively referred to as histocompatibility antigens.

HLA (Human leucocyte antigens) Proteins found on the membranes of nearly every nucleated cell in the body. They are the major determinants used by the body's immune system for recognition and differentiation of self from non-self.

Human immunodeficiency virus (HIV) A retrovirus that infects human $CD4^+$ cells and causes AIDS.

Humanization Genetic engineering of an antibody in which mouse CDR region genes defining a particular antigenic specificity are combined with human genes coding for the rest of the Ig molecule. This results in an antibody with a mouse-defined antigenic specificity but with human effector characteristics. Humanization reduces the immunogenicity of a mouse Ig in humans, making it more effective as an *in vivo* treatment than a completely mouse molecule.

Humanized antibody Genetically engineered mouse antibody in which the Fc (and sometimes the framework) regions have been replaced with human sequences to prevent a HAMA response.

Humira *See* Adalimumab.

Humoral immunity Refers to host defence resulting from the presence of specific antibody. This immunity can be obtained by transferring from one individual to another with serum.

Hybridoma An immortalized hybrid cell that results from the *in vitro* fusion of an antibody-secreting cell with a malignant cell; their clones secrete antibody without stimulation and proliferate continuously both *in vivo* and *in vitro.*

Hypersensitivity A state of reactivity to antigen that is greater than that of normal for the antigenic challenge; the exaggerated immune response causes deleterious outcome rather than a protective one.

Hypervariable regions Portions of the light and heavy immunoglobulin chains that show variability from one antibody to the other.

Ibritumonab tiuxetan *See* Zevalin.

ICAM Intercellular adhesion molecule that allows prolonged cell–cell contact during leucocyte activation or migration.

Idiotype The combined antigenic determinants (idiotopes) expressed in the variable region of antibodies of an individual, which are formed after binding with a particular antigen.

IFNα Alpha interferon; a cytokine produced in response to a viral infection and interferes with viral replication.

Ig *See* immunoglobulin.

IgA Alpha immunoglobulin found in mucus, body fluids and secretions. It is responsible for innate immunity. It occurs as a dimer.

IgD Delta immunoglobulin seen on the surface of B cells as a part of membrane–receptor complex; it induces signal transduction.

IgE Epsilon immunoglobulin found in serum or on the surface of mast cells.

IgG Gamma immunoglobulin found in the blood and responsible for secondary immune reaction.

IgM μ immunoglobulin found in the blood and responsible for primary immune response. It also occurs as a pentamer.

Immune complex Non-covalently bound complex of antibody with antigen, and sometimes complement.

Immunity The general term for resistance to a pathogen.

Immunization A process of response to antigen or vaccine.

Immunodeficiency Decrease in immune response resulting from absence or defect of some component of the immune system.

Immuno flow cytometry *See* FACS.

Immunofluorescence A detection of cell-associated molecules in the UV-fluorescent microscope with antibodies labelled with fluorochromes.

Immunogen A substance capable of inducing an immune response (as well as reacting with the products of an immune response). *See* antigen.

Immunoglobulin (Ig) A general term for all antibody molecules. Each Ig unit is made up of two heavy chains and two light chains and has two antigen-binding sites.

Immunohistochemistry Detection of cell-associated molecules in the light microscope with antibodies labelled with enzymes which change a substrate into a coloured precipitate.

Immunoliposome assay Detecting antibody is attached to a liposome, into which have been entrapped many detectable molecules such as enzymes or fluorescent or chemiluminescent reagents.

Immuno-PCR PCR utilizing monoclonal antibodies coupled to DNA through the avidin–biotin system.

Immunosensor A solid-state device in which the antigen–antibody reaction is detected via a transducer which provides an electric signal during the reaction.

Immunosuppression The suppression of the immune response. It is necessary following the organ transplants from a genetically different donor in order to prevent the host rejecting the grafted organ.

Immunotherapy The concept of using the immune system to treat a disease. Immunotherapy may also refer to the therapy of diseases caused by the immune system, e.g. allergies.

Immunotoxin A genetically engineered monoclonal antibody specific for a tumour cell. The Fc region has been modified or replaced with a cytotoxin (such as ricin or diphtheria toxin) which kills the tumour cell.

***In vitro* culturing** Culturing in an artificial environment, i.e., outside a living organism or body.

***In vivo* culturing** Culturing within a living organism or body.

Infliximab *See* Remicade.

Infusion A treatment in which a device (a drip) is attached to a patient, which constantly delivers a liquid into the bloodstream.

Interferon A group of glycoproteins secreted by certain cells, having antiviral activity and capable of enhancing and modifying the immune response.

Interleukins (IL) Glycoproteins secreted by a variety of leucocytes which have effects on growth and differentiation of other leucocytes.

Intrabodies Antibodies that are directed against intracellular target molecules and expressed within a specific subcellular compartment as directed by the intracellular localization signals genetically fused to N- or C-terminus of a given antibody.

Isotype An immunoglobulin molecule classified according to the amino acid sequences of the constant regions of heavy chains and/or light chains.

J chain (joining chain) A polypeptide involved in the polymerization of immunoglobulin molecules of IgM and IgA.

Knock-out mice A mice in which one particular gene has been inactivated; used to study gene function by looking at the effects of its absence.

Leukaemia An uncontrolled proliferation of a malignant leucocyte.

Light chain (L chain) A structural feature of immunoglobulin which occurs in two forms, kappa and lambda.

Lipopolysaccharide (LPS) *See* endotoxin.

Lymphocyte A small mononuclear cell virtually within the cytoplasm, found in blood, in all tissues, and in lymphoid organs, such as lymph nodes, spleen, and Peyer's patches, and bears antigen-specific receptors. They have specific receptors for antigen and participate in adaptive immunity.

Lymphokines Soluble substances secreted by lymphocytes, which have a variety of effects on lymphocytes and other cell types.

M13 A single-stranded DNA bacteriophage used as a vector for DNA sequencing.

Mass-detecting immunosensors Piezoelectric biosensors where quartz crystals are coated with antibody. Binding of an antigen results in a change in mass which is monitored through the change in the frequency of the crystal oscillation.

Microtitre plates Polypropylene or polycarbonate plates having 6 or 24 or 96 or 384 impressions or wells for the culturing, preservation and handling of cells.

Mixed lymphocyte reaction (MLR) A proliferative response occurring when leucocytes from two individuals are mixed *in vitro*; T cells from an individual (the responder) are activated by MHC antigens expressed by APC of the other individual (the stimulator).

Monoclonal Literally means coming from a single clone, i.e., the progeny of a single cell. In immunology, monoclonal generally describes an antibody that is monospecific.

Monoclonal 2F8 A second fully human monoclonal directed against EGFR.

Monoclonal antibody (mAb) Antibody produced from a clone of B cells. It is homogeneous; every molecule is identical in structure, isotype and antigen-binding specificity.

Monoclonal BL-22 It is used to treat hairy cell leukaemia with a protein on their surface called CD22. It is a genetically engineered immunotoxin conjugated with truncated *Pseudomonas* exotoxin 38.

Multiple myeloma A malignancy of plasma cells (a form of lymphocyte) that typically involves multiple sites within the bone morrow and secretes all or part of a monoclonal antibody. Also called plasma cell myeloma.

Multiple sclerosis A disease of the central nervous system believed to be autoimmune in nature, in which an inflammatory response results in demyelination and loss of neurologic function.

Muronomab *See* Orthoclone-OKT3.

Myeloma cell A tumour plasma cell generally secreting a single monoclonal immunoglobulin.

Myeloma protein The monoclonal antibody produced by myeloma cells.

Mylotarg A recombinant humanized (98.3%) IgG4 kappa antibody, conjugated with cytotoxic anti-tumour antibiotic calicheamicin isolated from fermentation of a bacterium, *Micromonospora echinospora* ssp. *calichensis*.

Natalizumab (Tysabri) A bio-engineered monoclonal antibody from part of a mouse antibody to closely resemble a human antibody for the treatment of patients with relapsing forms of multiple sclerosis to reduce the frequency of symptom flare-ups or exacerbations of the disease.

Neutropenia A pathological condition in which there is a decrease in the number of neutrophils.

NK cell Naturally occurring, large, granular, lymphocyte-like killer cells that kill various tumour cells. Also, they participate in ADCC. They are non-specific and their number does not increase by immunization.

Non-Hodgkin's lymphoma (NHL) A cancer of the lymph glands.

Nutraceuticals A food or a part of a food that provides medical or health benefits including the prevention and treatment of a disease.

Omalizumab A very high-affinity recombinant humanized anti-IgE monoclonal, which reduces symptoms of asthma, hay fever and other allergies.

Optical immunosensors Labelled antibody immunosensors based on fluorescence or luminescence. It is attractive due to the relative simple instrument design.

Orthoclone-OKT3 An IgG2a murine monoclonal binds to all T cells resulting in the early activation of T cells, which lead to cytokine release, followed by blocking T cell functions. The first FDA-approved immunosuppressive pharmaceutical with less side effects than other immunosuppressive agents for treating acute organ transplant rejection.

Osteoporosis A disease in which bones become fragile and are more likely to break.

Palivizumab The first monoclonal antibody successfully developed to combat an infectious disease caused by respiratory syncytial virus (RSV), a ubiquitous and highly contagious virus.

Panitumumab A monoclonal used for the treatment of EGFR-positive cancers.

Particulate antigen The antigens present on the cell or mounted on to some particles.

Passive immunization Immunization of an individual using pre-formed antibody from another individual.

Patent A set of exclusive rights granted by a government to an inventor or his assignee for a fixed period of time in exchange for a disclosure of an invention or a technique.

PEG (Polyethylene glycol) A fusing agent used to fuse cells.

Pegylated antibodies PEG-attached antibodies that are proteolytic-resistant and that have an increased circulating half-life.

Pep bodies Fusion of small peptides bound to natural immunoglobulin effector ligands or with antibody fragments such as Fab or scFv.

Percutaneous coronary intervention (PCI) A procedure to open blocked arteries of the heart.

Pharmaceutical agent *See* therapeutic agent.

Phytopharming The production of protein biologicals in recombinant plant systems.

Plantibody An antibody expressed transgenically in an engineered plant.

Plasma cell (end-stage differentiation of a B cell) The antibody-producing cell.

Plasmin and fibrin Activated proteins responsible for the degradation of clot.

Polyclonal activation Activation of B cells or T cells with several different antigenic specificities.

Polyclonal activator A substance that induces activation of many individual clones of either T cells or B cells.

Polyclonal antibody Antibody molecules with several different antigen-binding specificities.

Polyclonal antiserum A serum sample that contains a mixture of distinct immunoglobulin molecules, each recognizing a different antigenic determinant of a given antigen.

Precipitation An immunological assay for antibody–antigen complexes that is insoluble.

Primary cell culture The culturing of cells or cell lines taken directly from the living organism, which is not immortalized.

Primary response The immune response that occurs during the first encounter of immune system with a given antigen. The primary response is generally for a short time. It has a long induction phase or lag period, consists primarily of IgM antibodies, and generates immunologic memory.

Promoter The region of DNA where the enzyme RNA polymerase binds.

Protein kinase C Cytoplasmic enzyme that phosphorylates and activates transcription factor NFkB. It is the second messenger in lymphocyte inactivation.

Psoriasis A disease in which the cells of the skin reproduce more rapidly than normal, and appear scaly.

Radioimmunoassay (RIA) A widely used technique for measuring the primary antigen–antibody interactions, and for determining the level of important biological substances in mixed samples. It takes advantage of the specificity of the antigen–antibody interaction and the sensitivity that derives from the measurement of radioactively labelled materials.

Raptiva *See* Efalizumab.

Recombinant DNA DNA composed of DNA from more than one source.

Remicade The monoclonal antibody anti-TNF-α, to induce

temporary remission of rheumatoid arthritis, colitis, etc.

Reopro *See* Abciximab.

Repertoire The complete library of antigenic specificities generated by either B or T lymphocytes to respond to a foreign antigen.

Reverse transcriptase DNA-synthesizing enzyme which uses an RNA template and produces complementary DNA.

Rheumatoid arthritis An autoimmune, inflammatory disease of the joints.

Rhogam A fully human antibody to Rh antigen given to women to prevent haemolytic disease of the newborn.

RIA *See* Radioimmunoassay.

Rituximab (Rituxan or Mabthera) A monoclonal drug directed against CD20, a signature antigen of B cells, to treat non-Hodgkin's lymphoma.

ScFv Single chain fragment variable having variable regions of light and heavy chains connected by a linker peptide extending from C-terminal region of one chain to N-terminal region of another chain.

SCID mice Mice with severe combined immune deficiency resulting from the failure to develop mature T and B lymphocytes; suitable for adoptive transfer experiments in which their immune systems can be restored with normal cells.

Second-generation antibody therapies Therapies using modified monoclonals like glycosylated monoclonals, isotype-altered monoclonals, mutation-induced improved monoclonals, and domain-deleted monoclonals, scFv, Fab, etc.

Secondary response Immune response to the second exposure to antigen. It is quick and responded by T and B (memory) cells.

Secretory component A surface receptor on the epithelial cells lining mucosal surfaces which binds dimeric IgA and transports it through the cell into mucosal secretions.

Secretory IgA IgA which is secreted from mucosal epithelial cells in the digestive, respiratory, and genital tract. Secretory IgA has an additional protein (secretory component) that protects it from protease digestion.

Septic shock A serious condition caused by a large number of bacteria getting into blood. It is also called septicaemia, sepsis or blood poisoning.

Serotype Antigenic specificity of a pathogen.

Serum The liquid part of the clotted blood without the clotting factors.

Signal transduction Enzymatic cascade which follows ligand binding to receptor and results in change in cell function. Processes involved in transmitting the signal received on the outer surface of the cell (e.g. by antigen binding to its receptor) into the nucleus of the cell, which lead to altered gene expression.

Simulect A chimeric recombinant monoclonal that retains the murine variable region. It is a monoclonal against a trimeric cell-surface receptor of IFNα.

Spleen A secondary lymphoid organ located in the abdomen of the human body, where it destroys the old red blood cells. It consists of masses of lymphoid tissue of granular appearance located around fine terminal branches of veins and arteries.

Spondyloarthropathy An inflammatory condition affecting the spine and occasionally other joints. The condition is often characterized by back pain but the severity of the symptoms can vary greatly.

Synagis *See* Palivizumab.

Syngene(t)ic Genetically identical.

Systemic lupus erythematosis Systemic autoimmune disease characterized by facial rash (wolf-like markings, hence "lupus"), high levels of anti-DNA antibodies, and joint and kidney damage from immune complexes.

TAC Therapeutic Antibody Centre, Headington, Oxford.

Tetramas (quadromas) When two hybridomas fuse, it results in the formation of tetramas.

Titre The reciprocal of the last dilution of a titration giving a measurable effect, e.g. if the last dilution giving significant agglutination is 1 : 128, the titre is 128.

TNFα (Tumour-necrosis factor alpha) A cytokine produced by macrophages, monocytes and mast cells. Association of these monomers into a homotrimer yields active form.

TNFβ (Tumour-necrosis factor beta) A cytokine produced by T cells.

Tolerance The inability of the immune system to respond to an antigen.

Tositumomab *See* Bexxar.

Toxic residues Traces of toxic metabolites remaining in or on any commodity, animal, plant or environmental component. The term may be applied to contaminants of natural, industrial or environmental origin.

Transfectomas Hybridomas resulting from transfection of B cells with Epstein–Barr Virus.

Transformation Transfer of a recombinant plasmid vector into a bacterial cell, or conversion of a normal cell into a cancer cell.

Transgene Cloned foreign gene inserted into a cell.

Transgenic animals Genetically engineered animals, that are bred by a technique in which all or parts of genes from one animal are inserted experimentally into the genes of the embryos of another.

Transgenic mice Mice which have a foreign gene in their cells. Since the gene is inserted into the embryo it is present in all cells of the mouse but may not be expressed in all cells.

Transgenic plants Genetically modified plants.

Transgenic seeds Seeds that express transgenes.

Transplantation Grafting a solid tissue (such as a kidney or heart) or cells (particularly bone marrow) from one individual to another.

Trastuzumab *See* Herceptin.

Triomas When a hybridoma cell is fused with another B cell, it results in triomas.

Ulcerative colitis An inflammatory disease of the large bowel. It is not an allergic disease because it is not mediated by IgE, but has an immunologic basis.

Vitaxin A monoclonal antibody that binds to vascular integrin ($\alpha v_{\beta}3$) found on the blood vessels of tumours but not on the blood vessels supplying to normal tissues.

Western blot Polyacrylamide gel electrophoresis of proteins followed by blotting on nitrocellulose paper and binding of enzyme-tagged specific antibodies; used to detect specific antigens.

Xolair A recombinant DNA-derived humanized IgG1. It blocks IgE's high-affinity Fc receptor. It stops free-serum IgE from attaching to mast cells and other immune cells and prevents IgE-mediated inflammatory changes.

Zenapax A humanized monoclonal against IL2 receptor (anti-Tac) that retains only hypervariable region of murine antibody developed to reduce the immunogenicity.

Zevalin A murine monoclonal with radioactive isotope ^{111}Indium or ^{90}Y. It is used for the treatment of patients with the relapsed or refractory low-grade follicular non-Hodgkin's lymphoma. It binds to the CD20 antigen on B lymphocytes and induces apoptosis.

Index

Q

R

S

T

Made in the USA
Las Vegas, NV
07 June 2024

90850213R00155